T0264659

Controlling
SALMONELLA
in Poultry Production
and Processing

Controlling
SALMONELLA
in Poultry Production and Processing

SCOTT M. RUSSELL, PH.D.

CRC Press
Taylor & Francis Group
Boca Raton London New York

CRC Press is an imprint of the
Taylor & Francis Group, an **informa** business

CRC Press
Taylor & Francis Group
6000 Broken Sound Parkway NW, Suite 300
Boca Raton, FL 33487-2742

First issued in paperback 2016

ISBN 13: 978-1-138-19916-3 (pbk)
ISBN 13: 978-1-4398-2110-7 (hbk)

Library of Congress Cataloging-in-Publication Data

Russell, Scott M.
　　Controlling salmonella in poultry production and processing / Scott M. Russell.
　　　　p. ; cm.
　　"CRC title."
　　Includes bibliographical references and index.
　　Summary: "Salmonella is major pathogen that can result in foodborne illness. This book explains the origin of Salmonella on poultry and offers intervention strategies for controlling Salmonella during breeding, hatching, grow-out, transportation, and processing. The text examines the implications and proper use of chemicals, how to diagnose and tune a processing plant to eliminate Salmonella, and how to verify intervention strategies to ensure that they are working. It also discusses the implications of recycling water on Salmonella contamination and offers practical tips for increasing yield during processing while controlling for Salmonella and designing the proper equipment to eliminate Salmonella"--Provided by publisher.
　　　　ISBN 978-1-4398-2110-7 (hardcover : alk. paper)
　　　　I. Title.
　　　　[DNLM: 1. Salmonella Food Poisoning--prevention & control. 2. Food Contamination--prevention & control. 3. Food Handling. 4. Poultry. 5. Salmonella--pathogenicity. WC 268]

　　615.9'529
　　344--dc23　　　　　　　　　　　　　　　　　　　　　　　　　　　　　　　　　　　2011036604

Visit the Taylor & Francis Web site at
http://www.taylorandfrancis.com

and the CRC Press Web site at
http://www.crcpress.com

Contents

Acknowledgments

I would like to express my appreciation to God for the ability to complete this work; to my wife Kristine and my sons Jordan and Joshua for their support during this undertaking; and to my major professors, Dr. Daniel L. Fletcher and Dr. Nelson A. Cox, for the inspiration and scientific knowledge imparted to me during my time with them. I would also like to acknowledge the significant body of work by scientists with the United States Department of Agriculture (USDA) Agricultural Research Service in Athens, Georgia. Much of their work has been referenced in this book, and without the work contributed by these scientists, our understanding of this important topic would certainly be diminished.

Introduction

Salmonella causes considerable problems for the poultry industry in the United States each year. Poultry companies are required to control *Salmonella* on their raw and fully cooked products. The United States Department of Agriculture (USDA) Food Safety Inspection Service (FSIS) requires that each year inspectors sample poultry carcasses and send the samples to be tested for the presence of *Salmonella*. Plants that do not meet the requirement are penalized; therefore, the significance of this single bacterial genus is immense for poultry producers. The purpose of this book is to describe sources of *Salmonella* on poultry during breeding, hatching, grow out, and processing and to elucidate methods for controlling it during these processes.

About the Author

Dr. Scott M. Russell is a professor in the Poultry Science Department at the University of Georgia in Athens. He received his bachelor, master of science, and doctor of philosophy degrees from the university. He has worked in the department since 1994. Prior to that, he held positions as microbiologist, quality assurance manager, and production manager at Dutch Quality House in Gainesville, Georgia, and was head microbiologist at GoldKist Research Center, Lithonia, Georgia.

Dr. Russell's research activities have been primarily directed toward intervention strategies for reducing pathogenic and spoilage bacteria from poultry production and processing operations, and developing rapid microbiological methods for identifying and enumerating spoilage, indicator, and pathogenic bacteria from fresh and cooked poultry products. His research has resulted in a total of 44 refereed journal articles, 37 abstracts, 43 proceedings, 1 patent, 7 book chapters, and 71 popular articles for a total of 203 publications. He has been invited to speak 174 times at scientific meetings around the world. Dr. Russell has been featured on Fox News (*The Fox Report* with Shepard Smith) and *Good Morning America* (interviewed by Dr. Richard Besser, former director of the Centers for Disease Control and Prevention).

Dr. Russell works closely with the poultry industry throughout the United States and Canada and with countries in Central and South America, Europe, and China. He assists poultry companies with elimination of pathogenic bacterial populations throughout their grow-out and processing operations. In addition, Dr. Russell conducts applied research projects to assist in answering a variety of questions related to problems in poultry plants.

Chapter 1

Salmonella:
The Organism

1.1 Introduction

Salmonella is a significant problem in poultry throughout the world. Most governments regulate its presence on poultry, and the poultry industry in each of these countries is required to meet a "*Salmonella* standard." No other pathogen is so tightly regulated on poultry products.

1.2 Discovery and Origin of the Name

Salmonella was originally discovered by a technician named Theobald Smith in 1885; however, it was named after the technician's research leader, Daniel Salmon, who was a veterinarian. Daniel Salmon later became the founding director of the Bureau of Agriculture under the Department of Agriculture (Salmon and Smith, 1884–1886).

Salmonella is the genus name for a bacterium that is responsible for causing illness worldwide. Species in the genus *Salmonella* are categorized as facultatively anaerobic Gram-negative rods within the family Enterobacteriaceae. Most *Salmonella* are able to move (are motile) using peritrichous flagella distributed uniformly over the surface of each bacterial cell (Figure 1.1), except for *S. pullorum* and *S. gallinarum*, which do not possess flagella (Holt et al., 1994). *Salmonella* species grow best at

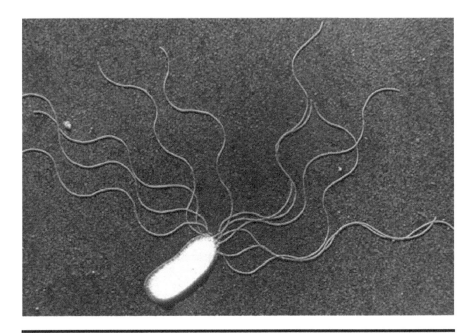

Figure 1.1   Peritrichous flagella on *Salmonella*. (Reproduced by permission from Cuppels D A, and Kelman A, 1980, Isolation of pectolytic fluorescent pseudomonads from soil and potatoes, *Phytopathology*, 70, 1110–1115.)

temperatures between 35°C and 42°C. For comparison, human body temperature is 37°C and a chicken's body temperature is 41.8°C, both well within the optimal temperature range of growth for *Salmonella*. The maximum growth temperature for *Salmonella* is 46.96°C (116.53°F; Juneja et al., 2009). Thus, *Salmonella* grow well in the intestines of both humans and poultry species. *Salmonella* are able to ferment carbohydrates into by-products such as acid and gas, and they use citrate as their sole carbon source. *Salmonella* produce H_2S as a by-product, do not produce the enzyme oxidase, and are able to produce the enzyme catalase (Lund et al., 2000).

1.3   Phylogenetic Characterization and Serotyping

For many decades, scientists have sought to separate *Salmonella* based on a number of phenotypic and genotypic characteristics in an effort to track the source of an outbreak. Salmonellae possess three major antigens used to categorize them into different serotypes; the H or flagellar antigen, the O or somatic antigen, and the Vi or capsular antigen (Southern Illinois University Carbondale). Salmonellae also possess the lipopolysaccharide (LPS) endotoxin characteristic of Gram-negative bacteria. This LPS is composed of an O polysaccharide (O antigen), an R core, and the endotoxic inner lipid A. These characteristics are important because endotoxins evoke fever in infected people and can activate significant immune responses including complement, kinin, and clotting factors (Southern Illinois University Carbondale).

Salmonella are unusual in that, unlike most other bacteria that are listed by their genus and species names, such as *Escherichia coli*, *Salmonella* are more often characterized by serotype. As early as 1938, scientists have sought to separate *Salmonella* strains even further within a particular serotype. The serotype may be separated into a number of phage types by their patterns of susceptibility to lysis by a series of bacteriophages with different specificities for different strains of *Salmonella*. The determination of the phage type of strains isolated from different patients, carriers, or other sources is valuable in the epidemiological study of infections as it helps to link outbreaks. By identifying groups of people who have been infected with the same strain from the same source, the source of an outbreak may be discovered. A high correlation exists between the phage type and the epidemic source of a *Salmonella* outbreak. The phage-typing method has become a well-established procedure in the routine epidemiological investigation of typhoid fever. Serotypes such as *S. typhi*, *S. paratyphi* A, *S. paratyphi* B, *S. typhimurium*, and *S. enteritidis* can be subdivided by phage typing. Approximately 106 different phage types of *S. typhi* and 232 different phage types of *S. typhimurium* have been distinguished using this method (Southern Illinois University Carbondale).

In 1972, Ewing identified three main species of *Salmonella*: *typhi*, *enteritidis*, and *choleraesuis*. As of the year 2000, *Salmonella* were separated into two distinct species: *enterica* and *bongori* (Lund et al., 2000). *Salmonella enterica* was then divided into six subspecies (*enterica*, *salamae*, *arizonae*, *diarizonae*, *houtenae*, and *indica*). In 2000, Lund et al. reported that there were approximately 2,422 name-bearing serovars of *Salmonella* belonging to the species *enterica*. There are additional serovars in the subspecies *enterica* and *bongori* that are not named but are identified by their antigenic formulae (Table 1.1) (Lund et al., 2000). Currently, most research publications use the genus name *Salmonella* and the serovar name (i.e., *kentucky*) to identify the bacterium.

Pulsed-field gel electrophoresis (PFGE) has been used to further separate *Salmonella* subtypes. The way that PFGE works is that, if the *Salmonella* cell is broken down and its DNA is removed, electrophoresis is used to move the DNA through a gel. The larger DNA fragments are less mobile than the smaller fragments as the fragments move through the gel. A threshold length exists above 30 to 50 kb where all large fragments will run at the same rate and appear in a gel as a single large diffuse band. However, with PFGE, there is a periodic changing of field direction, which causes the various lengths of DNA to react to the change at differing rates (Schwartz and Cantor, 1984). Larger pieces of DNA will be slower to realign their charge when field direction is changed, while smaller pieces will be more rapid. Schwartz and Cantor (1984) reported that over the course of time with the consistent changing of directions, each band will begin to separate increasingly

Table 1.1   Distribution of Serovars within the *Salmonella* **Genus**

Species	Subspecies	Number of Serovars
S. enterica	enterica	1,427
	salamae	482
	arizonae	94
	diarizonae	319
	houtenae	69
	indica	11
S. bongori		20
Total		2,422

Source: From Lund B M, Baird-Parker T C, and Gould G W, 2000, in *The Microbiological Safety and Quality of Food*, Volume 2, Aspen, Gaithersburg, MD, 1233–1299.

even at very large lengths. Thus, separation of very large DNA pieces using PFGE is possible. The PFGE procedure is similar to standard gel electrophoresis except that instead of constantly running the voltage through the gel in one direction, the voltage is periodically switched among three directions: one that runs through the central axis of the gel and two that run at an angle of 120° on either side (Schwartz and Cantor, 1984). Using PFGE, scientists have been able to further discriminate subtypes of *Salmonella* in poultry.

Fakhr et al. (2005) reported that PFGE is currently considered the "gold standard" technique in typing *Salmonella*. These researchers conducted studies to determine the discriminatory power of PFGE when compared to multilocus sequence typing (MLST) for typing *Salmonella enterica* serovar Typhimurium clinical isolates. MLST directly measures the DNA sequence variations in a set of housekeeping genes and characterizes strains by their unique allelic profiles. The principle of MLST is simple: The technique involves polymerase chain reaction (PCR) amplification followed by DNA sequencing. Differences in the nucleotides between strains can be checked at a variable number of genes (generally seven) depending on the degree of discrimination desired (http://en.wikipedia.org/wiki/Multilocus_sequence_typing). Fakhr et al. (2005) conducted a study with 85 *Salmonella* Typhimurium clinical isolates from cattle. The authors found that using MLST lacked the discriminatory power of PFGE for typing *Salmonella enterica* serovar Typhimurium.

Boxrud et al. (2007) reported that current subtyping methods yielded less-than-optimal subtype discrimination. The authors developed and evaluated a multiple-locus variable-number tandem repeat analysis (MLVA) method for sub-typing *Salmonella* serotype Enteritidis. In this study, the discrimination ability and epidemiological concordance of MLVA were compared to the traditional PFGE method and phage typing. Boxrud et al. (2007) found that MLVA provided greater discrimination among nonepidemiologically linked *Salmonella* Enteritidis isolates than did PFGE or phage typing. Epidemiologic concordance was evaluated by typing 40 isolates from four food-borne disease outbreaks. MLVA, PFGE, and, to a lesser extent, phage typing exhibited consistent subtypes within an outbreak. MLVA was better able to differentiate isolates between the individual outbreaks than either PFGE or phage typing (Boxrud et al., 2007). The reproducibility of MLVA was evaluated by subtyping sequential isolates from an infected individual and by testing isolates following multiple passages and freeze-thaw cycles. PFGE and MLVA patterns were reproducible for isolates that were frozen and passaged multiple times. However, 2 of 12 sequential isolates obtained from an individual over the course of 36 days had an MLVA type that differed at one locus, and one isolate had a different phage type. Overall, the authors found that MLVA typing of *Salmonella* serotype Enteritidis showed enhanced resolution, good reproducibility, and good epidemiological concordance (Boxrud et al., 2007). Results from this study demonstrated that MLVA may be a useful tool for detection and investigation of outbreaks caused by *Salmonella* serotype Enteritidis.

Some researchers have lamented that separating and characterizing *Salmonella* to this level is "splitting hairs." However, these methods are necessary to identify sources of outbreaks rapidly and when developing interventions to reduce them on poultry. Different phenotypes and genotypes of *Salmonella* may vary with regard to their susceptibility to specific disinfectants.

1.4 Contracting a *Salmonella* Infection

Salmonella live in the intestinal tracts of humans and other animals, including birds. These bacteria are usually transmitted to people by consuming foods that have been exposed to animal feces (http://www.cdc.gov/nczved/divisions/dfbmd/diseases/salmonellosis). Contaminated foods cannot usually be distinguished visually from uncontaminated foods. Some contaminated foods are of animal origin, such as beef, poultry, milk, or eggs; however, any food, including fruits, vegetables, nuts, and chocolate, may become contaminated. *Salmonella* is easily killed by cooking foods thoroughly, but in many cases, a food may become contaminated by the hands of an infected food handler who did not wash his or her hands with soap after handling raw meat or using the restroom (http://cdc.gov). *Salmonella* may also be found in the feces of some pets, especially those with diarrhea, and people can become infected if they do not wash their hands after contact with pets or pet feces. Reptiles, such as turtles, lizards, and snakes, are particularly likely to harbor *Salmonella*. Moreover, many chicks and young birds carry *Salmonella* in their feces.

1.5 Salmonellosis

Salmonellosis is an infection with one of the *Salmonella* serotypes. Most people who become infected with *Salmonella* develop diarrhea, fever, and abdominal cramps within 12 to 72 hours after consumption of the organisms (http://www.cdc.gov). The disease condition begins when a person ingests the bacterium, and the organism then colonizes the lower intestine. *Salmonella* are capable of invading the mucosa of the intestinal tract, which results in an acute inflammation of the mucosal cells (http://www.cdc.gov). The inflammation causes activation of adenylate cyclase, increased fluid production, and release of fluid into the intestinal lumen, resulting in diarrhea (Southern Illinois University Carbondale). The illness generally lasts 4 to 7 days, and most people will recover without any medical treatment. However, in some people, the diarrhea may be so severe that the patient needs to be hospitalized. In these patients, the *Salmonella* infection may spread from the intestines to the bloodstream and then to other body sites and can cause death unless the person is treated promptly with antibiotics. The elderly, infants, and those with impaired immune systems are more likely to have a severe illness (http://www.cdc.gov). People with diarrhea from salmonellosis usually recover completely, although it may be several months before their bowel habits are entirely normal. A small number of people with *Salmonella* develop pain in their joints, irritation of the eyes, and painful urination. This condition is called *Reiter's syndrome*. It can last for months or years and can lead to chronic arthritis, which is difficult to treat (http://www.cdc.gov). Antibiotic treatment does not make a difference in whether the person develops arthritis.

References

D'Aoust J Y (2000), Salmonella. In Lund B M, Baird-Parker A C, Gould G W, eds. *The Microbiological Safety and Quality of Food*, Vol. II, Chapter 45. Gaithersburg, MD: Aspen Publishers, Inc. 1233–1299.

Boxrud D, Pederson-Gulrud K, Wotton J, Medus C, Lyszkowicz E, Besser J, and Bartkus J M (2007), Comparison of multiple-locus variable-number tandem repeat analysis, pulsed-field gel electrophoresis, and phage typing for subtype analysis of *Salmonella* enterica serotype enteritidis, *Journal of Clinical Microbiology*, 45, 536–543.

Cuppels D A, and Kelman A (1980), Isolation of pectolytic fluorescent pseudomonads from soil and potatoes, *Phytopathology*, 70, 1110–1115.

Fakhr, M K, Nolan L K, and Logue C M (2005), Multilocus sequence typing lacks the discriminatory ability of pulsed-field gel electrophoresis for typing *Salmonella* enterica serovar Typhimurium, *Journal of Clinical Microbiology*, 43, 2215–2219.

Holt J G, Krieg N R, and Sneath P H A (1994), Genus *Salmonella*, in *Bergeys Manual of Determinative Bacteriology*, eds Holt J G, Krieg N R, and Sneath P H A, Williams and Wilkins, Baltimore, MD, 186–187.

Juneja V K, Melendres M V, Huang L, Subbiah J, and Thippareddi H (2009), Mathematical modeling of growth of *Salmonella* in raw ground beef under isothermal conditions from 10 to 45°C, *International Journal of Food Microbiology*, 131, 106–111.

Salmon D E, and Smith T (1884–1886), On a new method of producing immunity from contagious diseases, *Proceedings of the Biological Society in Washington,* 3, 29–33.

Schwartz D C, and Cantor C R (1984), Separation of yeast chromosome-sized DNAs by pulsed field gradient gel electrophoresis, *Cell,* 37, 67–75.

Chapter 2

The Social Cost of *Salmonella* Infections

2.1 Introduction

Food-borne salmonellosis constitutes a major health problem in many countries (Persson and Jendteg, 1992). During and immediately after World War II, salmonellosis first emerged as a public health problem in Britain, having been introduced primarily via contaminated batches of dried egg from the United States. In 1990, surveys of ready-to-cook broiler carcasses at retail outlets and hospitals have shown *Salmonella* contamination rates varying between 45% and 80% (Sharp, 1990). However, in the calendar year 2010, the U.S. Department of Agriculture, Food Safety Inspection Service (USDA-FSIS, 2010) analyzed 29,734 verification samples across eight meat and poultry product classes and found that only 6.7% of ready-to-cook broiler chickens, just prior to packaging, were positive for *Salmonella*. Thus, there has been a dramatic reduction in *Salmonella* prevalence on broilers from 1990 to 2010 in the United States.

The costs associated with human salmonellosis infections are considerable. Because of these costs, strong arguments exist for preventing this bacterium from entering the poultry production-and-processing system. Persson and Jendteg (1992) stated that government-sponsored programs aimed at preventing and controlling salmonellosis in poultry production represent one alternative to assist in lowering salmonellosis-related illness and economic costs. On the other hand, such comprehensive programs are resource demanding (Persson and Jendteg, 1992).

Food-borne diseases cause approximately 76 million illnesses, 325,000 hospitalizations, and 5,000 deaths in the United States each year (Mead et al., 1999). Rostagno et al. (2006) stated that *Salmonella* is the second most common cause of bacterial food-borne diseases, and poultry products are implicated as a major source of human food-borne salmonellosis. During slaughter and processing, *Salmonella* from the gastrointestinal tract of carrier birds can contaminate carcasses and the slaughter and processing line. Because of these concerns, in 1996, the USDA-FSIS published the pathogen reduction/HACCP (hazard analysis and critical control point) final rule (USDA-FSIS, 2001). This rule required that poultry companies in the United States control *Salmonella* and that the FSIS begin testing poultry carcasses in all plants to determine *Salmonella* prevalence. The reasons given by FSIS regarding why it considered *Salmonella* to be the appropriate organism to use as the measure of performance in pathogen reduction include the following:

1. *Salmonella* is a problem pathogen that is among the most common causes of food-borne illnesses associated with meat and poultry products.
2. *Salmonella* is relatively easy to find using current testing methodologies.
3. *Salmonella* is a useful indicator, meaning that interventions aimed at reducing *Salmonella* are likely to be beneficial in reducing contamination by other enteric pathogens.
4. It is relatively easy to monitor *Salmonella* because it occurs at frequencies that permit changes in its occurrence to be detected.

FSIS chose *Salmonella* as its target because it felt that it would provide a clear indication of whether sanitation standard operating procedures (SSOPs) and HACCP systems were succeeding in controlling and reducing pathogens (USDA-FSIS, 2001). Therefore, *Salmonella* in poultry is considered an extremely important problem to the FSIS, the poultry industry, and consumers.

2.2 Calculating the Cost of Infection

Various agencies have attempted to estimate the social cost of salmonellosis to U.S. citizens. In 1989, Todd reported that microbiological diseases (bacterial and viral) represent 84% of U.S. foodborne costs, with salmonellosis one of the two most widespread and economically important diseases. The authors estimated the socioeconomic cost to be $4.0 billion based on approximately 2.9 million cases annually, and this affects all sectors of the food industry. Later, the estimates were reduced, with Bryan and Doyle (1995) reporting that estimates place the annual incidence of human salmonellosis in the United States at approximately 1 million cases. The authors noted that annual costs, including lost work time and medical care of poultry-associated cases of salmonellosis, in the United States range from

Table 2.1   Estimation of the Costs Associated with *Salmonella* Infections in the United States by USDA-FSIS

Step in Calculation	Input	Information for Salmonella	Data Source
1	Incidence	14.4/100,000	FoodNet Annual Report for 2003
2	Population estimate 2003	290,788,976	U.S. Census Bureau
3	Underreporting multiplier	38	Mead et al.
4	Food-borne fraction	0.95	Mead et al.
5	Poultry attribution factor	0.3351	Food Safety Research Consortium
6	Young poultry fraction	0.838	ERS
7	Total illnesses	1,591,197	Step = (1)(2)(3)
8	Total food-borne illnesses	1,511,637	Step = (4)(7)
9	Total food-borne illnesses from poultry	498,840	Step = (5)(8)
10	Total food-borne illnesses from young chickens	424,389	Step = (6)(9)
11	Costs per illness	$1,800	ERS
12	Total costs of illnesses from product and pathogen	$759,000,000	Step = (10)(11)

$64 million to $114.5 million. Voetsch et al. (2004), using a model they developed, estimated that 1.4 million people contract nontyphoidal *Salmonella* infections in the United States each year, resulting in 168,000 physician office visits per year during the years 1996–1999. Including both culture-confirmed infections and those not confirmed by culture, the authors estimated that *Salmonella* infections resulted in 15,000 hospitalizations and 400 deaths annually (Voetsch et al., 2004).

In 2007, Engeljohn of the USDA-FSIS presented a table describing how cost estimates for *Salmonella* infections in the United States are calculated. The information is presented in Table 2.1.

In 2008, *Morbidity and Mortality Weekly Report* ("Preliminary FoodNet Data," 2008) reported that in 2007 a total of 6,790 cases of *Salmonella* were confirmed, which equates to 14.92 cases/100,000 people, consistent with the USDA-FSIS estimate. These calculations indicate that the annual cost of *Salmonella* infections in the United States from poultry is enormous, at $759 million; however, this may be refuted, as is discussed in Chapter 29.

2.3 Reasons for Underreporting Salmonellosis

Voetsch et al. (2004) stated that there are many reasons why salmonellosis may be underreported. First, a person infected with *Salmonella* must develop symptoms that are severe enough for him or her to seek medical care. Second, the physician must request and collect a specimen from the patient and forward it to a microbiology laboratory for bacterial culture. In many cases, physicians do not request culturing, but instead use a broad-spectrum antibiotic as a "shotgun" approach. Third, the laboratory must test the specimen appropriately for *Salmonella* using a sensitive method and, if *Salmonella* is identified, forward this isolate to a state public health laboratory for serotyping. Fourth, the state laboratory, in turn, must report the serotype result to the Centers for Disease Control and Prevention (CDC). Although about 30,000 to 40,000 culture-confirmed cases of nontyphoidal *Salmonella* are reported to the CDC each year through the national surveillance system, these cases have been estimated to represent only 1–5% of the actual number of nontyphoidal *Salmonella* infections that occur. Thus, it is difficult to estimate the total number of cases of salmonellosis each year.

2.4 Cost to the U.S. Poultry Industry

Aside from the social costs associated with *Salmonella* infections, there is a cost that the U.S. poultry industry must pay to comply with USDA-FSIS regulations. According to the USDA-FSIS (1996), Table 2.2 shows the estimate of costs to the industry to remain in compliance with the HACCP and *Salmonella* performance standard implemented by USDA in 1996.

Table 2.2 Summary of Annual Industry Costs to Comply with the USDA-FSIS *Salmonella* **Performance Standard**

Cost Category	Year 1	Year 2	Year 3	Year 4	Year 5
Compliance with *Salmonella* standards	No estimate	$5,472,000–$16,899,000	$5,353,000–$25,753,000	$5,811,000–$25,956,000	$5,811,000–$26,079,000
Cumulative total expense					$22,447,000–$94,687,000

The cost is estimated at $5,472,000 to $16,899,000 in Year 2, increasing until Year 5 (2001) on, which has an estimate of $5,811,000 to $26,079,000. These cost estimates are quite low in that many plants spend a minimum of $500,000 per year just for the online reprocessing (OLR) chemicals. Multiplying 160 processing plants by $500,000 per year gives a figure of $80 million (more than three times the maximum cost listed in the table). It should be noted that not all plants use an OLR system, but the costs are far more than calculated. The 5-year cumulative total is estimated at $22 to $95 million. These figures do not include chemicals or interventions used during breeding, hatching, grow out, scalding, online applications other than OLR, chilling, or postchill dips. The costs also do not address new processing equipment that many companies have had to purchase to achieve acceptable *Salmonella* levels on their products, additional labor required to collect and test samples for *Salmonella*, or the capital cost of the microbiological testing equipment required for testing. *Salmonella* prevention and intervention on poultry are expensive and becoming increasingly important as the USDA-FSIS implements new and more difficult standards. Hence, the need for a cost-effective solution to these poultry-borne human disease problems is apparent.

References

Bryan F L, and Doyle M P (1995), Health risks and consequences of *Salmonella* and *Campylobacter jejuni* in raw poultry, *Journal of Food Protection*, 58, 326–344.

Engeljohn, E (2007), Expert elicitation and its role in RBI, United States Department of Agriculture, Food Safety Inspection Service. http://www.fsis.usda.gov/PDF/RBI_062607_Engeljohn.pdf.

FoodNet Annual Reports (2003), http://www.cdc.gov/foodnet/reports.htm.

Mead P S, Slutsker L, Dietz V, Mccaig L F, Bresee J S, and Shapiro C (1999), Food-related illness and death in the United States, *Emerging Infectious Diseases*, 5, 607–625.

Persson U, and Jendteg S (1992), The economic impact of poultry-borne salmonellosis: How much should be spent on prophylaxis? *International Journal of Food Microbiology*, 15, 207–213.

Preliminary FoodNet data on the incidence of infection with pathogens transmitted commonly through food—10 states (2008), *Morbidity and Mortality Weekly Report*, 57, 366–370. http://www.cdc.gov/mmwr/preview/mmwrhtml/mm5714a2.htm.

Rostagno M H, Wesley I V, Trampel D W, and Hurd H S (2006), *Salmonella* prevalence in market-age turkeys on-farm and at slaughter, *Poultry Science*, 85, 1838–1842.

Sharp J C M (1990), Salmonellosis, *British Food Journal*, 92, 6–12.

Todd E C D (1989), Preliminary estimates of costs of food-borne disease in the United States, *Journal of Food Protection*, 52, 595–601.

U.S. Department of Agriculture, Food Safety Inspection Service (1996), Pathogen reduction; hazard analysis and critical control point (HACCP) systems, *Federal Register*, 61, 38806–38989.

U.S. Department of Agriculture, Food Safety Inspection Service (2001), Performance standards for *Salmonella* on carcasses and on raw ground product. http://www.fsis.usda.gov/OPPDE/rdad/FRPubs/01–013N/Derfler.htm.

U.S. Department of Agriculture, Food Safety Inspection Service (2010), Progress report on *Salmonella* testing of raw meat and poultry products, 1998–2010. http://www.fsis. usda.gov/science/progress_report_salmonella_testing/index.asp.

Voetsch A C, Van Gilder T J, Angulo F J, Farley M M, Shallow S, Marcus R, Cieslak P R, Deneen V C, and Tauxe R V (2004), FoodNet estimate of the burden of illness caused by nontyphoidal *Salmonella* infections in the United States for the Emerging Infections Program FoodNet Working Group, Foodborne and Diarrheal Diseases Branch, Division of Bacterial and Mycotic Diseases, Centers for Disease Control and Prevention, and Emory University School of Medicine, Atlanta, GA; California Emerging Infections Program, San Francisco, CA; Yale University, New Haven, CT; Office for Disease Prevention and Epidemiology, Oregon Department of Human Services, Portland, OR; and Minnesota Department of Health, Minneapolis, MN, *Clinical Infectious Diseases* (Suppl. 3), S127–S134.

Chapter 3

Risk Assessment of *Salmonella* from Poultry Sources

3.1 Introduction

In the United States, the authority for food safety oversight is divided between the Food Safety and Inspection Service (FSIS) of the U.S. Department of Agriculture (USDA) and the Food and Drug Administration (FDA) of the U.S. Department of Health and Human Services, as reported by Guo et al. (2007). The FSIS regulates the production of meat, poultry, and egg products under the authority of the Meat Inspection Act, the Poultry Products Inspection Act, and the Egg Products Inspection Act. The FDA regulates other foods under the authority of the Federal Food, Drug, and Cosmetics Act.

Attributing food-borne illness outbreaks to specific food types can assist risk managers and policy makers in formulating public health goals, prioritizing interventions, and documenting the effectiveness of prevention efforts for reducing illness and improving public health. To assess risk, the USDA-FSIS adapted a Bayesian statistical model to quantify attribution of meat, poultry, and eggs as sources of human salmonellosis in the United States (Guo et al., 2007). Assessment of food product safety and attribution of food-borne illnesses requires extensive data, originating from a number of different sources.

3.2 Statistical Model Used for Risk Analysis

Hald et al. (2007) described a statistical model that combines epidemiologic surveillance data, pathogen prevalence data, and food consumption data. The model has been used to attribute cases of human salmonellosis in Denmark to specific food commodities and has led to the implementation of food commodity-specific policies that have reduced the incidence of food-borne salmonellosis. Human infections caused by *Salmonella* subtypes found in multiple animal reservoirs are attributed proportionally to the occurrence of each of the specific subtypes. Microbial subtyping provides a link between the source of infection and human food-borne illnesses. The Danish *Salmonella* model uses a Bayesian framework that applies Markov chain Monte Carlo simulation to estimate the expected number of human *Salmonella* infections. The approach quantifies the contribution of each of the major animal-food sources to human salmonellosis.

Hald et al. (2007) called for other countries to apply their model to *Salmonella* surveillance data to promote integration of quantitative risk assessment and zoonotic disease surveillance. Guo et al. (2007) conducted such a risk assessment.

The key equation used by Hald et al. (2007) in the Danish *Salmonella* attribution model was $l_{ij} = p_{ij}M_ja_jq_i$, where l_{ij} is the expected number of cases for a particular serotype i and product j, p_{ij} is the prevalence of *Salmonella* serotype i in product j, M_j is the amount of product j available for consumption, a_j is the food product-dependent factor for product j, and q_i is the bacteria-dependent factor for serotype i. Attribution of salmonellosis cases to the specific food was estimated using data for (1) the number of observed human salmonellosis cases; (2) the prevalence of *Salmonella* serotypes in seven food commodities (ground beef, intact beef, chicken, turkey, pork, shell eggs, and egg products); and (3) human consumption of these food commodities. The probable values of a_j and q_i were determined using Bayesian inference given observational data for the total number of salmonellosis cases of each *Salmonella* serotype and the prevalence of the serotypes in the seven food commodities.

3.3 Adaptation of the Risk Model

Next, the Danish model of attribution had to be adapted to be used with data from the United States. In the adapted model, as in the original version, a simplistic assumption was made that all of the human cases of salmonellosis addressed by the model were associated with a defined set of food commodities, either directly by the patient consuming that food or indirectly through the commodity's contamination of other foods (Guo et al., 2007).

The data on human cases used in the adapted model were all cases of salmonellosis reported through the Public Health Laboratory Information System (PHLIS) during the 6-year period 1998–2003. This reporting system covers all 50 states of the United States. For each of the FSIS-regulated commodities, steers and heifers (i.e., young fed cattle), cows and bulls (i.e., older cattle), ground beef, ground turkey, pork, broiler chickens, and pasteurized egg products, *Salmonella* test results of the hazard analysis and critical control point (HACCP) monitoring program were used to estimate the prevalence of *Salmonella*, by serotype, in each of product categories except for shell eggs during the 6-year period 1998–2003. Shell egg data were taken from the Pennsylvania *Salmonella* Enteritidis Pilot Project, 1993–1995 (Guo et al., 2007).

Per capita food consumption data were obtained from the Food Consumption Data System of the Economic Research Service (http://www.ers.usda.gov/Data/FoodConsumption/). Yearly data for 1998 through 2003 were obtained for each of the seven food categories included in this study. For U.S. data, *Salmonella* serotypes differ in their likelihood of association with outbreaks and with travel. Because this model was designed to focus on the attribution of sporadic disease, the initial task was to estimate the number of human cases, by serotype, that were due to domestically acquired sporadic cases.

3.4 Attribution Data

Figure 3.1 shows the preliminary results of relative percentage of estimated culture-confirmed salmonellosis cases from intact beef, ground beef, chicken, turkey, pork, eggs, egg products, and other sources for the period 1998–2003 in the United States. Of all cases of salmonellosis estimated for this period, the largest proportion of cases (19% of all cases) was attributed to ground beef, followed by those attributed to chicken (18% of all cases), eggs (12% of all cases), turkey (8% of all cases), pork (2% of all cases), intact beef (<1% of all cases), and egg products (<1% of all cases). About 41% of all cases of salmonellosis for this period were not attributed to any of the seven food categories in the model (Guo et al., 2007).

These data indicate that 38% of food-borne infections of *Salmonella* were due to consumption of a poultry product (chicken, turkey, or eggs). However, it should be noted that 41% of illnesses were due to other foods, such as vegetables and nuts, and there are currently no laws regarding testing and controlling *Salmonella* on these products. These foods represent a huge risk to consumers.

3.5 Expert Elicitation of Public Health Risks

Guo et al. (2007) reported that the FSIS recently conducted an expert elicitation to rank the public health risks posed by bacterial hazards and to attribute food-borne illnesses to specific pathogens as a result of consuming or handling processed meat

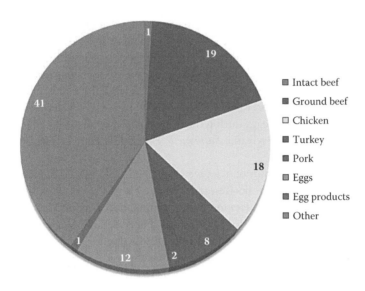

Figure 3.1 Estimated percentage distributions of human salmonellosis cases, 1998–2003. (From Guo C, Schroeder C, and Kause J, 2007, *Challenges in data needs for assessment of food product risk and attribution of foodborne illnesses to food products in the United States,* **Food Safety and Inspection Service, Office of Public Health Science, Washington, DC. http://www.stats.gov.cn/english/ICAS/ papers/P020071114301170316028.pdf.)**

and poultry products. The expert elicitation ranked the public health risks posed by bacterial hazards in each of the 25 categories of processed meat and poultry products for healthy adults and for vulnerable consumers, respectively. It also ranked the confidence level on a scale of 1 to 3, with 1 indicating "little or no confidence," and 3 indicating "very confident." The results of the expert elucidation are presented in Tables 3.1 and 3.2.

The results of the 2007 expert elicitation-attribution of food-borne illness of *Salmonella* to meat and poultry products are presented in Figure 3.2.

The researchers found that to assess food product risk and attribute illnesses to food products adequately, they needed to obtain data concerning pathogen prevalence and distribution in a wide variety of potential food vehicles and for other important sources of human exposure, such as indirect sources of contamination and nonfood sources. Another variable that was encountered was the need to ensure that existing data sources continue to adequately represent the burden of food-borne illnesses in the U.S. population and the distribution of the associated pathogen in food vehicles and exposure sources of interest. Finally, Guo et al. (2007) expressed the need to refine existing data so that the comparisons between data from various sources were based on similar units of observation at the necessary levels of discrimination for defined points along the farm-to-table continuum.

Table 3.1 Top Seven Product Types and Their Likelihood of Causing Illness among Healthy Adults as a Result of Consuming or Handling Finished Product Types

Finished Product Type	Median Score (1–10)	Level of Confidence (1–3)
Raw ground or otherwise nonintact chicken	10	2.6
Raw ground or otherwise nonintact turkey	9	2.3
Raw ground or otherwise nonintact, not chicken or turkey	8.5	1.8
Raw intact chicken	8	2.6
Raw intact turkey	8	2.5
Raw intact poultry, other than chicken or turkey	8	1.9

Source: From Guo C, Schroeder C, and Kause J, 2007, Challenges in data needs for assessment of food product risk and attribution of foodborne illnesses to food products in the United States, Food Safety and Inspection Service, Office of Public Health Science, Washington, DC. http://www.stats.gov.cn/english/ICAS/papers/P020071114301170316028.pdf.

Table 3.2 Top Seven Product Categories and the Likelihood of Causing Illness among Vulnerable Consumers as a Result of Consuming or Handling Finished Product Types

Finished Product Type	Median Score (1–10)	Level of Confidence (1–3)
Raw ground or otherwise nonintact chicken	10	2.6
Raw ground or otherwise nonintact beef	9.5	2.5
Raw ground or otherwise nonintact turkey	9	2.5
Raw ground or otherwise nonintact, not chicken or turkey	9	2.0
Raw intact chicken	8.5	2.6
Raw intact turkey	8	2.6
Raw intact poultry, other than chicken or turkey	8	2.1

Source: From Guo C, Schroeder C, and Kause J, 2007, Challenges in data needs for assessment of food product risk and attribution of foodborne illnesses to food products in the United States, Food Safety and Inspection Service, Office of Public Health Science, Washington, DC. http://www.stats.gov.cn/english/ICAS/papers/P020071114301170316028.pdf.

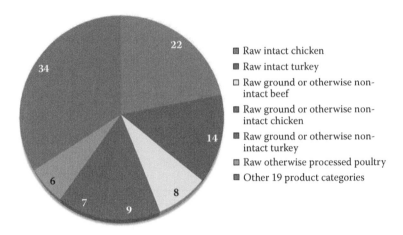

Figure 3.2 Attribution of food-borne illness of *Salmonella* to meat and poultry products. (From Guo C, Schroeder C, and Kause J, 2007, Challenges in data needs for assessment of food product risk and attribution of foodborne illnesses to food products in the United States, Food Safety and Inspection Service, Office of Public Health Science, Washington, DC. http://www.stats.gov.cn/english/ICAS/papers/P020071114301170316028.pdf.)

3.6 European Food Safety Authority Findings

In 2008, the European Food Safety Authority (EFSA) published, "Quantitative Microbiological Risk Assessment on *Salmonella* in Meat: Source Attribution for Human Salmonellosis from Meat," which was based on a scientific opinion of the Panel on Biological Hazards in Europe (Scientific Opinion of the Panel, 2008). Food vehicles linked to outbreaks of *Salmonella* have been summarized previously by D'Aoust (2000), Scientific Committee on Veterinary Measures Related to Public Health (2003), O'Brien et al. (2006), and Hughes et al. (2007). Eggs, egg products, broiler meat, and some red meat, especially pork, are consistently identified as a source for *Salmonella* in meat food-borne outbreaks of salmonellosis. Of 3,406 *Salmonella* outbreaks reported in the European Union, meat products were only implicated as vehicles 5.3% of the time, with a total of 179 occurrences, but in the largest category (meat and offal unspecified) the animal origin of meat/offal implicated was unknown (Table 3.3).

In two reviews of 1,426 food-borne general outbreaks of infectious intestinal diseases in England and Wales between 1992 and 1999, 20% were associated with the consumption of poultry (Kessel et al., 2001), and 16% were linked with the consumption of red meat (Smerdon et al., 2001). For the poultry-associated outbreaks, chicken was implicated in almost three-quarters of these outbreaks, turkey in over

Table 3.3 *Salmonella* Outbreaks Reported in the European Union in 2005 Related to Meat and Meat Products

Animal Species	Number of Reported Outbreaks
Meat and offal unspecified	**78**
Unspecified	54
Hot dog	1
Salami	1
Mixed meat	2
Mixed meat product	1
Minced meat	6
Minced meatballs	1
Raw meat	2
BBQ	1
Kebab	8
Liver	1
Broiler/chicken	**69**
Unspecified	45
Roast	10
Product	5
Kebab	2
Soup	2
Pepper chicken	2
Nuggets	1
Breasts	1
Chicken and bowels	1
Turkey	**12**
Unspecified	9
Roast	1

continued

Table 3.3 (continued)   *Salmonella* Outbreaks Reported in the European Union in 2005 Related to Meat and Meat Products

Animal Species	Number of Reported Outbreaks
Cutlets	1
Sausage	1
Pig	**11**
Unspecified	8
Meat preparation	1
Shashlik	1
Roast hog	1
Beef	**6**
Unspecified	3
Steak	1
Raw/carpaccio/tartare	2
Lamb	**2**
Duck	**1**
Total	**179**

Source: Scientific Opinion of the Panel, 2008, A quantitative microbiological risk assessment on *Salmonella* in meat: *Source* attribution for human salmonellosis from meat, *European Food Safety Authority Journal,* 625, 1–32.

Note: Bold indicates total.

a fifth, and duck in 2% of outbreaks. The organisms most frequently reported were *Salmonella* (30% of outbreaks), *Clostridium perfringens* (21%), and *Campylobacter* (6%). In these reviews, over 7,000 people were affected, with 258 hospital admissions and 17 deaths. In the red meat–associated outbreaks, over 5,000 people were affected, with 186 hospital admissions and 9 deaths. Beef (34%) and pig meat (32%) were the most frequently implicated red meat types, with lamb implicated in 11% of outbreaks. *Salmonella* was the second most frequently identified organism in these outbreaks (34.3%).

3.7 Estimated Risk Associated with Specific Meat Types in England and Wales

Adak et al. (2005) used data from outbreaks to attribute food-borne disease to its source in England and Wales. Table 3.4 shows the risks associated with a variety of meat types. The authors reported that the most important cause of U.K.-acquired food-borne disease was contaminated chicken (398,420 cases, risk = 111, case-fatality rate = 35, deaths = 141).

Red meat (beef, lamb, and pork) contributed heavily to deaths, even though there were lower levels of risk associated with its consumption (287,485 cases, risk = 24, case-fatality rate = 57, deaths = 164). The authors made note that, in these analyses, it was impossible to determine whether the contaminated meat had been ground up.

Table 3.4 Estimated Risks Associated with Types of Meat, England and Wales, 1996–2000 for All Pathogens

Food Group/Type	Disease Risk	Risk Ratio	Hospitalization Risk	Risk Ratio
Poultry	**104**	**947**	**2,063**	**4,584**
Chicken	111	1,013	2,518	5,595
Turkey	157	1,429	645	1,433
Mixed/unspecified	24	217	852	1,893
Red meat	**24**	**217**	**102**	**227**
Beef	41	375	153	339
Pork	20	180	93	208
Bacon/ham	8	75	39	86
Lamb	38	343	128	285

Source: Adak G K, Meakins S M, Yip H, Lopman B A, and O'Brien S J, 2005, Disease risks from foods, England and Wales, 1996–2000, *Emerging Infectious Diseases*, 11, 365–372.

Note: Bold indicates total.

3.8 The Risk of Extrapolating Information from Outbreak Datasets

The EFSA opinion (Scientific Opinion of the Panel, 2008) reported that extrapolating information from outbreak datasets in an attempt to describe food-borne *Salmonella* burden is not straightforward. A major limitation is investigation bias. Large outbreaks, outbreaks associated with food service and institutions, and outbreaks that have short incubation times or cause serious disease are more likely to be investigated and reported (O'Brien et al., 2002). As a result of this bias, the data may not reflect what occurs in sporadic cases. Another major limitation identified by the committee is that it is assumed that the relative pathogen-specific contribution of each food type to both sporadic and outbreak-associated disease is similar and, therefore, that outbreak experience can be generalized to sporadic disease. However, certain vehicles may be more likely to be implicated in outbreaks than others, especially if investigators preferentially collect data on the types of food perceived as high risk or when laboratory methods vary in sensitivity according to food type. An excellent example of this situation is when a person goes to a doctor with a food-borne illness and symptoms that are consistent with salmonellosis. The doctor will often ask, "When was the last time you had chicken to eat?" This type of approach significantly biases data that are gathered for risk assessments.

The EFSA opinion (Scientific Opinion of the Panel, 2008) also found that a third limitation is that, in many outbreaks, it is not possible to find an etiological agent or identify a source of infection. In a detailed overview of *Salmonella* outbreaks published by D'Aoust (2000), he found that the published outbreaks represent a biased fraction of all outbreaks.

3.9 Microbial Subtyping

As mentioned, microbial subtyping is extensively used for tracking outbreaks to source and to identify diffuse outbreaks but has been applied as an attribution method only in the Netherlands (Van Pelt et al., 1999) and in Denmark (Hald et al., 2004). Although the basic idea behind the two methods is similar, the approaches differ with regard to the statistical methods applied and the number of parameters in the model. The Dutch approach compares the number of reported (domestically acquired, sporadic) human cases caused by a particular *Salmonella* type with the relative occurrence of that type in the animal-food sources. Results of attribution modeling for the Netherlands are shown in Table 3.5. Van Pelt et al. (1999) reported that throughout 1994–2005, eggs and pork were the two most important sources of human salmonellosis in The Netherlands, accounting for up to two-thirds of all cases in 2003.

Table 3.5 Estimated Contribution (%) of Different Reservoirs to Laboratory-Confirmed Salmonellosis in the Netherlands

Reservoir	1994–98	2001–2	2003	2004	2005	2006
Pig	24	25	26	23	24	21
Cattle	10	14	12	11	11	13
Chicken	19	15	11	13	14	14
Layers	37	35	37	37	32	36
Travel/other	9	11	13	15	19	16

Source: From Van Pelt W, Van De Giessen A W, Van Leeuwen W J, Wannet W, Henken A M, and Evers E G, 1999, Oorsprong, omvang en kosten van humane salmonellose, Deel 1. Oorsprong van humane salmonellose met betrekking tot varken, rund, kip, ei en overige bronnen, *Infectieziekten Bulletin,* 10, 240–243; and Valkenburgh S, Van Oosterom R, Stenvers O, Aalten M, Braks M, Schimmer B, Van De Giessen A W, and Langelaar M, 2007, *Zoonoses and Zoonotic Agents in Humans, Food, Animals and Feed in the Netherlands 2003–2006,* Centrum Infectieziektebestrijding, Bilthoven, The Netherlands.

Included in the EFSA opinion (2008) is Figure 3.3, depicting the estimated major sources of human salmonellosis in Denmark in 2005 (Anon., 2006).

In this study, regionally produced and imported poultry products were estimated to be responsible for 29.45% of *Salmonella* outbreaks in Denmark.

3.10 Trends in Attribution of Salmonellosis from Broilers and Table Eggs

In another study (Anon., 2006), trends in attribution were made from data for broilers and table eggs. The data are presented by year in Figure 3.4.

In another study in Denmark (Anon., 2006), human salmonellosis attributed to consumption of broiler chicken meat decreased dramatically from 1988 (1,600 cases) to 2005 (40 cases). However, salmonellosis attributed to consumption of contaminated eggs increased significantly from 1988 (300 cases) to 1997 (3,000 cases) and then decreased by 2005 (200 cases). The authors reported that the reason for these decreases by 2005 was that salmonellosis was found to decrease in response to interventions in the broiler meat chain (1988) and the egg chain (1997) (Wegener et al., 2003).

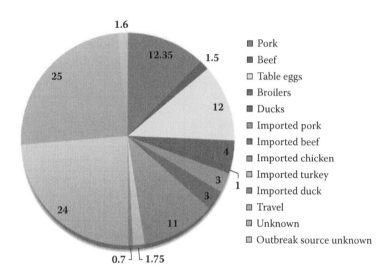

Figure 3.3 The major sources of human salmonellosis in Denmark in 2005. (From Anonymous, 2006, *Annual report on zoonoses in Denmark 2005*, Ministry of Family and Consumer Affairs, Copenhagen, Denmark. Available from http://www.food.dtu.dk.)

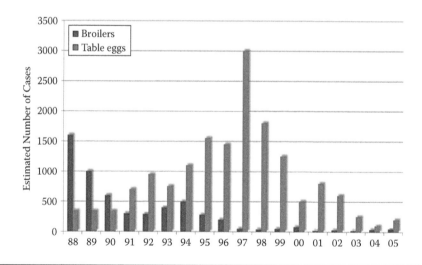

Figure 3.4 Trends in the attribution of major sources of human salmonellosis in Denmark 1988–2005. (From Anonymous, 2006, *Annual report on zoonoses in Denmark 2005*, Ministry of Family and Consumer Affairs, Copenhagen, Denmark. Available from http://www.food.dtu.dk.)

In a study in the United States (Hoffmann et al., 2006, 2007), 44 experts from different backgrounds (government, industry, academia) and different scientific disciplines (medicine, food science, public health, microbiology, and veterinary medicine) met to estimate food category attribution to a variety of food products. Expert estimates were compared with estimates based on outbreaks, as published previously on the basis of data from the Centers for Disease Control and Prevention (CDC). Data for *Salmonella* spp. are shown in Table 3.6. The experts considered poultry to be the main source of salmonellosis, whereas outbreak data suggested eggs to be the dominant source. Pork appears to be a relatively small source of salmonellosis in the United States, based on outbreak data and in particular on expert estimates.

In Table 3.7, the food category attribution data for salmonellosis in the Netherlands are presented.

Table 3.6   Food Category Attribution (% of cases) of Salmonellosis in the United States Based on Structured Expert Judgment and Outbreak Data

Food Category	Expert Estimate	Outbreak Data
Poultry	35	18
Eggs	22	37
Produce	12	17
Beef	11	6
Dairy	7	7
Pork	6	3
Seafood	2	—
Luncheon and other meats	2	—
Beverages	2	—
Game	2	—
Breads and bakery	<1	—

Source: Hoffmann S, Fischbeck P, Krupnick A, and McWilliams M, 2006, Eliciting information on uncertainty from heterogeneous expert panels attributing U.S. foodborne pathogen illness to food consumption, Resources for the Future, Washington, DC. http://www.rff.org/rff/Documents/RFF-DP-06–17.pdf; and Hoffmann S, Fischbeck P, Krupnick A, and McWilliams M, 2007, Using expert elicitation to link foodborne illnesses in the United States to foods, *Journal of Food Protection*, 70, 1220–1229.

Table 3.7   Food Category Attribution (% of cases) of Salmonellosis in the Netherlands Based on Structured Expert Judgment

Food Category	Expert Estimate	Range 5–95th Percentile
Eggs and egg products	22	11–54
Chicken meat and other poultry meat	15	5–47
Pork	14	6–36
Beef and lamb	13	5–28
Dairy products	7	0–25
Fruit and vegetables	6	0–20
Other foods, including composite foods	6	0–18
Infected humans and animals	6	0–18
Fish and shellfish	4	0–10
Bread, grains, pastas, and bakery products	4	0–12
Beverages	<1	—

Source: Vargas-Galindo, 2007.

3.11   Trend in Food-Borne Illness Due to *Salmonella* and the Serotypes Involved

Henao (2007) with the CDC presented a graph (Figure 3.5) depicting how *Salmonella* infections in people have changed in comparison to the baseline level established between 1996 and 1998.

It is interesting to note that, while *Salmonella* prevalence on raw poultry increased significantly between 2000 and 2005 and decreased dramatically between 2005 and 2010, no real impact on human salmonellosis was observed. This means that the introduction of the HACCP/*Salmonella* performance standard had a slight impact on relative rates of infection 4 years after its implementation, but the effect was short-lived. In 2002, even with the incredible expense associated with implementation and operation of the new inspection program, *Salmonella* infection rates in the United States were identical to the baseline 6 years after its introduction. Overall, very little impact has been made in human salmonellosis rates due to this performance standard, even though billions of dollars have been spent by the poultry industry to comply with this standard.

Figure 3.6 indicates that human infections with *Salmonella* Typhimurium have decreased steadily over the last decade. However, human infections with *Salmonella* Enteritidis have been variable, with an overall trend upward.

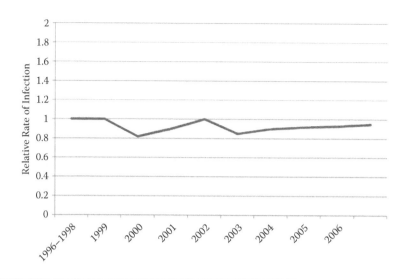

Figure 3.5 Relative rates compared with baseline data from 1996 to 1998 of laboratory-diagnosed cases of infection with *Salmonella* by year. (From Henao O, 2007, Foodborne diseases active surveillance network, Centers for Disease Control and Prevention, Atlanta, GA. http://www.fmi.org/foodsafety/presentations/CDC_Data-Henao-April_2007.pdf.)

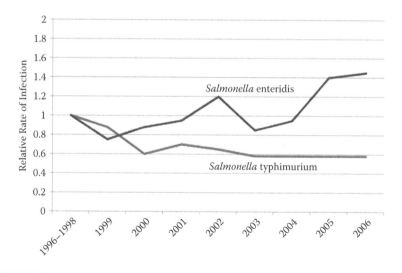

Figure 3.6 Relative rates compared with baseline data from 1996 to 1998 of laboratory-diagnosed cases of infection with *Salmonella typhimurium* and *Salmonella enteritidis* by year. (From Henao O, 2007, Foodborne diseases active surveillance network, Centers for Disease Control and Prevention, Atlanta, GA. http://www.fmi.org/foodsafety/presentations/CDC_Data-Henao-April_2007.pdf.).

3.12 Model to Assess the Risk of Acquiring Salmonellosis from Consumption or Handling of Chickens

In 1998, Oscar developed a simulation model to assess the risk of acquiring salmonellosis from consumption and handling of chickens. The model simulated the distribution, preparation, and consumption of 1,000 chickens and was designed to determine the relationship between the level of *Salmonella* contamination on chickens at the processing plant exit and the risk of salmonellosis for consumers of the chickens. Using a scatterplot of the probability of acquiring salmonellosis from consumption of the chickens simulated versus the *Salmonella* load on the chickens at the processing plant exit, the author was able to demonstrate that highly contaminated (i.e., >100 *Salmonella*/carcass) carcasses at the plant exit did not necessarily pose greater risk of salmonellosis when compared to carcasses that had low levels (i.e., <10 *Salmonella*/carcass) at the plant exit. Instead, Oscar (1998) found that a greater risk of salmonellosis was realized from carcasses with low levels of contamination when they were temperature abused, undercooked, and consumed by someone from the high-risk population. These findings shift the responsibility from the government inspection agencies and the poultry-processing companies to the shoulders of the consumer. This means that food preparers and consumers play a larger role in food safety than the government or processors.

3.13 Effect of the Finnish *Salmonella* Control Program

To study the public health effects of the Finnish *Salmonella* control program (FSCP), Maijalaa et al. (2005) developed a quantitative risk assessment model of *Salmonella* being transferred from slaughtered broiler flocks to consumers. Based on this model, in Finland, approximately 0.21% of domestically produced broiler meat was contaminated with *Salmonella* (95% probability interval 0.05–0.48%). Using this method, the effect on public health of eliminating breeder flocks from production that have tested positive for *Salmonella* and heat treating the meat of detected positive broiler flocks could be simulated. Based on the entire model, if detected positive breeder flocks were not removed, 1.0- to 2.5-fold more human cases would occur when compared to the expected number of cases under the current FSCP (95% predictive interval). Without heat treatment of meat, the increase would be 2.9- to 5.4-fold, and without both interventions, *Salmonella* infections in humans would increase 3.8- to 9.0-fold. The authors found that this model suggested that with a higher infection level, inclusion of both interventions would be more effective than either of the interventions alone. Replacement of half of the current retail broiler meat by meat with 20–40% contamination could result in 33 to 93 times more human cases compared to the expected value under current

Finnish regulations (Maijalaa et al., 2005). Thus, on the basis of this model, the interventions applied in FSCP significantly protect the public health.

3.14 *Salmonella* Enteritidis Surveillance

In 2006, Marcus et al. reported that active surveillance for laboratory-confirmed *Salmonella* serotype Enteritidis (SE) infection revealed a decline in incidence in the 1990s, followed by an increase starting in 2000. The authors conducted a population-based case-control study of sporadic SE infection in five of the Foodborne Diseases Active Surveillance Network (FoodNet) sites during a 12-month period in 2002–2003. A total of 218 cases and 742 controls were enrolled. Sixty-seven (31%) of the 218 case-patients and six (1%) of the 742 controls reported travel outside the United States during the 5 days before the illness onset for the case. Of SE phage type 4 cases, 81% traveled internationally. Among persons who did not travel internationally, eating chicken prepared outside the home and undercooked eggs inside the home were associated with SE infections. Contact with birds and reptiles was also associated with SE infections. Interestingly, chicken prepared inside the home were not associated with SE infections.

3.15 Proper Handling Labels

The U.S. National Academy of Sciences (NAS) committee recommended that warning labels be required on poultry slaughtered and sold in the United States to alert consumers to the possible risk from *Salmonella* and other bacteria present in more than 33% of poultry in the U.S. food system (Anon., 1987). The committee also urged that the labels describe proper cooking and handling procedures to avoid food poisoning from contaminated poultry. However, the NAS committee could not recommend specific steps for the poultry industry to take to reduce bacterial contamination. The committee recommended that the USDA adopt a new technique of risk assessment but would not support USDA's desire to eliminate federal inspectors that USDA feels are unnecessary since they cannot see bacterial contamination.

References

Adak G K, Meakins S M, Yip H, Lopman B A, and O'Brien S J (2005), Disease risks from foods, England and Wales, 1996–2000, *Emerging Infectious Diseases*, 11, 365–372.
Anonymous (1987), Warning labels needed on poultry; NAS report silent on industry role, *Nutrition Week*, 17, 1–2.

Anonymous (2006), Annual report on zoonoses in Denmark 2005, Ministry of Family and Consumer Affairs, Copenhagen, Denmark. http://www.food.dtu.dk.

D'Aoust J Y (1997), *Salmonella* in *Food Microbiology Fundamentals and Frontiers*, eds Doyle M P, Beuchat L R, and Montville T J, ASM Press, Washington, DC, 129–158.

D'Aoust J Y (2000), Salmonella. In Lund B M, Baird-Parker A C, Gould G W, eds. *The Microbiological Safety and Quality of Food*, Vol. II, Chapter 45. Gaithersburg, MD: Aspen Publishers, Inc., 1233–1299.

Guo C, Schroeder C, and Kause J (2007), Challenges in data needs for assessment of food product risk and attribution of foodborne illnesses to food products in the United States, Food Safety and Inspection Service, Office of Public Health Science, Washington, DC. http://www.stats.gov.cn/english/ICAS/papers/P020071114301170316028.pdf.

Hald T, Vose D, and Wegener H C (2004), Quantifying the contribution of animal-food sources to human salmonellosis by a Bayesian approach, *Risk Analysis,* 24, 255–269.

Hald T, Wong D M, and Aerestrup F M (2007), The attribution of human infections with antimicrobial resistant Salmonella bacteria in Denmark to sources of animal origin, *Foodborne Pathogenic Diseases*, 4, 313–326.

Henao O (2007), Foodborne diseases active surveillance network, Centers for Disease Control and Prevention, Atlanta, GA. http://www.fmi.org/foodsafety/presentations/CDC_Data-Henao-April_2007.pdf.

Hoffmann S, Fischbeck P, Krupnick A, and McWilliams M (2006), Eliciting information on uncertainty from heterogeneous expert panels attributing U.S. foodborne pathogen illness to food consumption, Resources for the Future, Washington, DC. http://www.rff.org/rff/Documents/RFF-DP-06–17.pdf.

Hoffmann S, Fischbeck P, Krupnick A, and McWilliams M (2007), Using expert elicitation to link foodborne illnesses in the United States to foods, *Journal of Food Protection*, 70, 1220–1229.

Hughes C, Gillespie I A, and O'Brien S J (2007), The breakdowns in food safety group, foodborne transmission of infectious intestinal disease in England and Wales, 1992–2003, *Food Control*, 18, 766–772.

Kessel A S, Gillespie I A, O'Brien S J, Adak G K, Humphrey T J, and Ward L R (2001), General outbreaks of infectious intestinal disease linked with poultry, England and Wales, 1992–1999, *Communicable Diseases and Public Health*, 4, 171–177.

Maijalaa R, Ranta J, Seuna E, Pelkonen S, and Johansson T (2005), A quantitative risk assessment of the public health impact of the Finnish *Salmonella* control program for broilers, *International Journal of Food Microbiology*, 102, 21–35.

Marcus R, Varma J K, Medus C, Boothe E J, Anderson B J, Crume T, Fullterton K E, Moore M R, White P L, Lyszkowicz E, Voetsch A C, and Angulo F J (2006), Re-assessment of risk factors for sporadic *Salmonella* serotype Enteritidis infections: A case-control study in five FoodNet sites, 2002–2003, *Epidemiology and Infection,* 1–9. http://www.cdc.gov/enterics/publications/182-marcus_2006.pdf.

O'Brien S J, Elson R, Gillespie I A, Adak G K, and Cowden J M (2002), Surveillance of food-borne outbreaks of infectious intestinal disease in England and Wales 1992–1999: Contributing to evidence-based food policy? *Public Health*, 116, 75–80.

O'Brien S J, Gillespie I A, Sivanesan M A, Elson R, Hughes C, and Adak C G (2006), Publication bias in foodborne outbreaks of infectious intestinal disease and its implications for evidence-based food policy in England and Wales 1992–2003, *Epidemiology and Infection*, 134, 667–674.

Oscar T P (1998), The development of a risk assessment model for use in the poultry industry, *Journal of Food Safety*, 18, 371–381.

Scientific Committee on Veterinary Measures Related to Public Health (2003), Scientific
opinion on salmonellae in foodstuffs, adopted 14–15 April 2003. European
Commission, Health Consumer Protection, Directorate-General.

Scientific Opinion of the Panel (2008), A quantitative microbiological risk assessment on
Salmonella in meat: Source attribution for human salmonellosis from meat, *European
Food Safety Authority Journal*, 625, 1–32.

Smerdon W J, Adak G K, O'Brien S J, Gillespie I A, and Reacher M (2001), General outbreaks of infectious intestinal disease linked with red meat, England and Wales, 1992–
1999, *Communicable Diseases and Public Health*, 4, 259–267.

Valkenburgh S, Van Oosterom R, Stenvers O, Aalten M, Braks M, Schimmer B, Van De
Giessen A W, and Langelaar M (2007), *Zoonoses and Zoonotic Agents in Humans, Food,
Animals and Feed in the Netherlands 2003–2006*, Centrum Infectieziektebestrijding,
Bilthoven, The Netherlands.

Van Pelt W, Van De Giessen A W, Van Leeuwen W J, Wannet W, Henken A M, and Evers E G
(1999), Oorsprong, omvang en kosten van humane salmonellose, Deel 1. Oorsprong
van humane salmonellose met betrekking tot varken, rund, kip, ei en overige bronnen,
Infectieziekten Bulletin, 10, 240–243.

Vargas-Galindo A (2003), Probabilistic inversion in priority setting of foodborne pathogens.
MSc. Thesis, Delft University of Technology, Department of Applied Mathematics
and Risk Analysis.

Wegener H C, Hald T, Wong D L F, Madsen M, Korsgaard H, Bager F, Gerner-Smidt P,
and Mølbak K (2003), *Salmonella* control programs in Denmark, *Emerging Infectious
Diseases,* 9, 774–780. http://www.cdc.gov/ncidod/EID/vol9no7/03–0024.htm.

Chapter 4

Sources of *Salmonella* in the Breeder Flocks, Hatchery, and Grow-Out Operations

4.1 Introduction

When considering a practical approach to preventing *Salmonella* from entering the preharvest poultry production system or eliminating *Salmonella* in the preharvest operation once it has been detected, companies have a difficult time because there are so many potential sources for the organism (Bryan and Doyle, 1995). These sources include the baby chicks (vertical transmission), feed, rodents or wild birds that enter the breeder or grow-out house, insects (including darkling beetles or flies), transportation coops, tractors or vehicles entering the grow-out house during clean out, and cows that graze near grow-out houses. Bailey et al. (2001) conducted a landmark study to characterize the potential sources of *Salmonella* regarding their relative level of importance. These authors identified variables that contribute to *Salmonella* contamination, such as (1) age of the chicken, (2) survival of the *Salmonella* through the gastric barrier, (3) competing bacteria in the intestinal tract, (4) availability of a hospitable colonization site, (5) the nature of the chicken's diet, (6) physiological status of the chicken, (7) health and disease status of the chicken, and (8) medication effects that influence the potential colonization of the chickens with *Salmonella* (Bailey et al., 2001). As far back as 1952, Milner and Shaffer discovered that the ability of *Salmonella* to colonize baby chicks was dependent on the amount of *Salmonella* they were exposed to, and at 1 day of age, chicks could be colonized by as few as five *Salmonella* cells, whereas after that time, colonization became irregular and required higher doses. Cox and others (1990) demonstrated that day-old chicks could be colonized with only two *Salmonella* cells if administered to their cloaca, as might occur in the hatchery or grow-out house if a baby chick were to sit down on contaminated litter. However, after the chick reaches 2 weeks of age, the chickens have competent gut flora and are much more resistant to colonization by *Salmonella* (Barnes et al., 1972).

4.2 Seasonal Variation

Bailey et al. (2001) reported that *Salmonella* prevalence on chickens varies by the season of the year (Figure 4.1). These data demonstrated that *Salmonella* prevalence in fall is far greater than summer.

4.3 Sources of *Salmonella* in Turkey Flocks

In turkeys, Hoover et al. (1997) conducted an ecological survey from March 1995 to February 1996 to determine the sources of *Salmonella* colonization in two flocks of turkeys reared consecutively in a newly constructed facility. Sampling was conducted prior to placement of poults, at Day 0, and again after 2, 10, 14, or 18 weeks. Samples were collected at comparable times for the second flock except that final sampling occurred after 22 weeks instead of 18 weeks. Samples included poult box liners, the

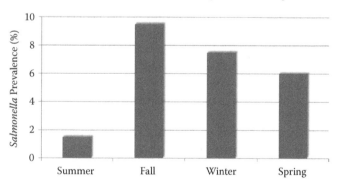

Figure 4.1 *Salmonella* **prevalence plotted by season of the year.**

birds themselves, new litter, drinkers, and air. Feed was collected from each truck-load as it arrived at the facility. Feeders, drinkers, and used litter were monitored to evaluate potential horizontal transmission. Before placement of the first flock of poults, the litter, drinkers, and air samples were all negative for the presence of *Salmonella*, whereas the drinkers were positive for *Salmonella* before the place-ment of the second flock. After poults were placed on the litter, 51.1%, 63.8%, and 22.8% of all litter, drinker, and air samples, respectively, became positive for *Salmonella* (Hoover et al., 1997). *Salmonella* was also isolated from 13.6% of the poult box liners, 25.0% of yolk sac samples, and 53.8% of ceca, excluding Day 0. Moreover, *Salmonella* was isolated from 14.8% of feed shipments and 39.1% of feeder contents. Frequency of *Salmonella* detection was higher ($p < 0.05$) in Flock 1 than Flock 2 for cecal and air samples. *Salmonella* colonization of turkey flocks and the spread of *Salmonella* within the environment were extensive once initial contamination of the production house occurred. Drinkers, feeders, litter, and air were critical sources of horizontal transmission within each pen as well as between pens (Hoover et al., 1997). In the following chapters, each of the processes, includ-ing breeding, hatching, and grow out, are examined in detail, and the sources of contamination within these operations are identified.

4.4 Vertical Transmission

Numerous investigators have implicated breeders as vehicles for vertical transmis-sion of *Salmonella* from the breeder chickens to the fertile egg. In 1991, Cox et al. evaluated egg fragments, paper pads from chick boxes, and chick fluff (from the bottom of the cabinet) samples from six commercial broiler breeder hatcheries for the presence and level of salmonellae. Overall, 42 of 380 samples (11.1%) from those hatcheries were contaminated with salmonellae. Salmonellae organisms were detected in 22 of 145 (15.2%), 5 of 100 (4.6%), and 15 of 125 (12%) samples of egg fragments, chick fluff, and paper pads, respectively. The percentage salmonellae-positive samples from each of the six hatcheries were 1.3%, 5.0%, 22.5%, 11.4%, 36.0%, and 4.3% (Cox et al., 1991). Of the 140 samples randomly selected for enu-meration, salmonellae were found in 11 samples. Four of these 11 samples had greater than 10^3 salmonellae per sample, 3 others had greater than 10^2 but less than 10^3, and the remaining 4 had less than 10^2. *Salmonella* serotypes isolated were *S. berta*, *S. california*, *S. give*, *S. hadar*, *S. mbandaka*, *S. senftenberg*, and *S. typhimurium*, all of which have previously been isolated from poultry. The authors found that the incidence and extent of salmonellae-positive samples in the breeder hatcheries were much less than that previously found in broiler hatcheries. Cox et al. (1991) concluded by stating that the cycle of salmonellae contamination will not likely be broken until contamination at these critical points is eliminated.

4.5 Feed Implicated

A survey of contamination with *Salmonella* was done in the breeder/multiplier and broiler houses, feed mills, hatcheries, and processing plants of two integrated broiler firms (Jones et al., 1991). Samples of insects and mice were also collected at each location. Of the meat and bone meal samples collected at feed mills, 60% were contaminated. *Salmonella* was isolated from 35% of the mash feed samples tested. The pelleting process was able to reduce *Salmonella* isolation contamination by 82.0%. In this study, Jones et al. (1991) concluded that feed was the ultimate source of *Salmonella* contamination in breeder houses. *Salmonella* was found in 9.4% of the yolk sac samples collected from day-old chicks in hatcheries coming from these breeders.

A retrospective, case-control study into risk factors of salmonellosis was undertaken using data from 111 broiler breeder flocks assembled during a 5-year period (Henken et al., 1992). Many different *Salmonella* serotypes were detected. The authors concluded that the following variables appeared to be the most relevant to determining whether birds would be contaminated with *Salmonella*: disinfection tubs, biosecurity, and feed mills. The final model indicated that flocks housed at farms without an egg disinfection tub, with poor hygiene barriers, and receiving their feed from a small feed mill had a 46.1 times greater risk of being *Salmonella* positive than flocks housed at farms with an egg disinfection tub, with good biosecurity, and if the breeder farm received its feed from a large feed mill.

4.6 Semen Implicated

In 1995, Reiber et al. conducted three experiments to determine the bacteriological quality of rooster semen. Semen was collected from donor males, diluted, and surface inoculated onto seven different bacteriological media, from which randomly selected colonies were identified. The most frequently isolated genera from rooster semen included *Escherichia*, *Staphylococcus*, *Micrococcus*, *Enterococcus*, and *Salmonella*. Most of the bacteria that were isolated were endemic to poultry and were commonly found in the environment of chickens (Reiber et al., 1995). Thus, during mating, female breeders may become inoculated with *Salmonella* during semen transmission.

4.7 Rodent Transmission Implication

In a study to determine the effect of disinfection on *Salmonella* in breeders, three broiler breeder houses at three different locations were sampled before and after cleansing and disinfection (Davies and Wray, 1996). None of the farms was able to achieve total elimination of *Salmonella* Enteritidis from the poultry house environment; however, substantial improvements were observed when errors in the cleansing and disinfection protocol in the first house had been corrected. Fundamental errors such as overdilution and inconsistent application of disinfectants were observed despite supervision of the process by technical advisors. The authors concluded that, in each of the three breeder houses, failure to eliminate mice from the house that was infected with *Salmonella* Enteritidis was likely to be the most important hazard for transmission to the next flock (Davies and Wray, 1996).

4.8 Eggshell Penetration Studies

Studies were conducted to determine how well *Salmonella* was able to penetrate the eggshell and membranes in hatching eggs from a commercial broiler breeder flock (Berrang et al., 1998). Figure 4.2 shows an electron micrograph of an egg shell pore. It is easy to see how *Salmonella* can penetrate the shell.

Egg weight, specific gravity, conductance, and ability of *Salmonella* to penetrate the shell and membranes were determined. Thirty unsanitized eggs were sampled on Weeks 29, 34, 39, 42, 48, 52, and 56 of flock age for specific gravity and conductance. An additional 10 intact eggs were inoculated with *Salmonella* using a temperature differential immersion method for 1 minute. Eggs were then emptied of contents and filled with a selective medium that allowed visualization of *Salmonella* growth on the inside of the shell and membrane complex. The authors reported that, over the 27-week sampling period, egg weight increased from 56 to 66 g and was positively correlated with hen age ($r = 0.96$, $p < 0.05$). However, neither specific gravity (ranging from 1.077 to 1.082) nor eggshell conductance (ranging from 14.7 to 17.9 mg weight loss/day) showed any clear trend throughout the life of the

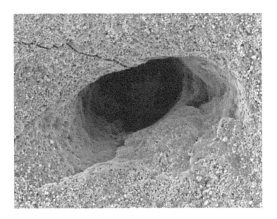

Figure 4.2   A pore in a chicken eggshell. (Used with permission from Jim Ekstrom. From http://cumberlandmuseum.net/jekstrom/SEM/SEM.html.)

flock, despite the increase in egg weight. Conductance values were not correlated with specific gravity. The number of eggs positive for *Salmonella* penetration after 24-hour incubation showed a general upward trend with flock age; however, penetration frequency and hen age were not significantly correlated ($p > 0.05$). There was no relationship between egg specific gravity, conductance, or egg weight and the likelihood of *Salmonella* to penetrate the eggshell. Because shell characteristics did not change over time and the penetration patterns did vary, it is likely that factors other than specific gravity and conductance were involved in the penetration of eggshells by *Salmonella*.

4.9 Breeder Implication in Vertical Transmission

Cox et al. (2000) reported that numerous publications showed that *Salmonella*-contaminated eggs can be produced by artificially inoculating breeding chickens (Figure 4.3). Timoney et al. (1989) reported that oral inoculation of laying hens resulted in infection of the reproductive tract. Challenging the breeder hen with 10^6 *Salmonella* cells caused the ovary and oviduct to become infected. Cox et al. (2000) observed that the egg production rate for infected chickens was unaffected,

Figure 4.3   Breeder chicken flock.

and *Salmonella* was not detected in all fecal samples; therefore, breeders infected with salmonellae may not always be easily detectable on the farm. For the contaminated breeder hens, the yolks of 10% of the eggs laid were contaminated with *S. enteritidis*. However, when hens were inoculated with *Salmonella* at levels of 10^8 cells, a noticeable drop in egg production and signs of pathogenesis occurred (Cox et al., 2000).

Shivaprasad et al. (1990) confirmed Timoney et al.'s (1989) findings that inoculation with lower numbers did not cause noticeable signs in the hens but resulted in contaminated eggs. Using different routes of inoculation, Miyamoto et al. (1997) examined the location of contamination in the oviduct. The authors found that intravenous inoculation caused colonization of the ovary and contamination of eggs while in the oviduct. When the inoculum was placed into the vagina, colonization of only the lower portions of the oviduct occurred, but eggs were produced with internal contamination. Therefore, some internal contamination of eggs may result from the lower oviduct and may actually be due to penetration of the shell in the oviduct, not colonization of the ovary.

In 1995, Keller et al. reported that lower oviduct contamination was important in the production of infected eggs. The authors discovered that while forming, eggs may be contaminated due to the colonization of an inoculated hen's ovary. This contamination sometimes decreases as the egg progresses through the oviduct; however, on entering a contaminated lower oviduct, the egg can then be recontaminated (Keller et al., 1995). These studies indicated that the egg is subject to challenge in both the upper and lower oviduct. Gast and Beard (1990) found that noninoculated hens can become contaminated and lay infected eggs just by being exposed to inoculated penmates. This fact raises the question of how birds may be contaminated under farm conditions and makes it difficult to distinguish between vertical and horizontal transmission of *Salmonella*. It has been shown that the hen's ovary can be colonized with *S. enteritidis* through airborne inoculation (Baskerville et al., 1992). Humphrey et al. (1992) suggested that *Salmonella* may enter the chicken through the conjunctiva when airborne *Salmonella* is prevalent. The authors found that delivery of about 100 cells to the eye of laying hens produced *Salmonella* infection of the ovary and oviduct. Thus, the reproductive tract of a laying hen can be colonized experimentally. Numerous reports detailed a variety of methods to detect low numbers of *Salmonella* that may be present in a few eggs (Gregory, 1948; Gast and Beard, 1992; Gast, 1993). When examining naturally contaminated hens, the prevalence of *S. enteritidis* in eggs is low. In 1992, Poppe and others observed that less than 0.065% of eggs tested (two positive samples from 16,000 eggs) were positive. In two separate studies, Humphrey et al. (1989, 1991) found very few *S. enteritidis*-contaminated eggs when sampling naturally contaminated flocks; however, these researchers also found eggs with other strains of *Salmonella*.

Sander et al. (2001) investigated *Salmonella* contamination at a U.S. commercial quail operation. Pulsed-field gel electrophoresis (PFGE) was used to type *Salmonella* isolates to trace them throughout this production environment. During a 6-month survey, *Salmonella* serotypes *hadar, typhimurium, typhimurium* variant Copenhagen, and *paratyphi* were encountered within this poultry operation. Ninety-four percent of the *Salmonella* isolated from breeder and production houses and from carcass rinses belonged to *Salmonella* serotypes *typhimurium* variant Copenhagen and *hadar*. There were six distinct *S. typhimurium* variant Copenhagen genetic types, as identified by PFGE, present within this particular poultry operation. Seventy-nine percent of *S. typhimurium* variant Copenhagen identified from the environment of the breeder and production houses produced the same PFGE pattern. Thirty-eight percent of *S. typhimurium* Copenhagen isolated from carcass rinses and the breeder house shared the same PFGE DNA pattern. The authors concluded that this study demonstrated the vertical transmission of salmonellae from the breeders to their progeny and to the birds ultimately processed for human consumption (Sander et al., 2001).

In 2002, Bailey et al. stated that, although the widespread presence of *Salmonella* in all phases of broiler chicken production and processing is well documented, little information is available to indicate the identity and movement of specific serotypes of *Salmonella* through the different phases of an integrated operation. In the study by Bailey et al. (2002), samples were collected from the breeder farm, the hatchery, the previous grow-out flock, the flock during grow out, and carcasses after processing. *Salmonella* were recovered from 6%, 98%, 24%, 60%, and 7% of the samples, respectively, in the first trial and from 7%, 98%, 26%, 22%, and 36% of the samples, respectively, in the second trial. Seven different *Salmonella* serotypes were identified in the first trial, and 12 different serotypes were identified in the second trial (Bailey et al., 2002). Interestingly, for both trials there was poor correlation between the serotypes found in the breeder farms and those found in the hatchery.

4.10 Implication of *Salmonella* in Breeder Follicles

Cox et al. (2005) conducted five trials to determine whether *Campylobacter* and *Salmonella* spp. existed naturally in the mature and immature ovarian follicles of late-life broiler breeder hens. Broiler breeder hens ranging from 60 to 66 weeks of age were obtained from four different commercial breeder operations. For each of these trials, the hens were removed from the commercial operation and held overnight. The hens were then euthanized, defeathered, and aseptically opened (Cox et al., 2005). To reduce the possibility of cross contamination between samples, first the mature and immature ovarian follicles (Figure 4.4), then the ceca, were aseptically removed. Samples were placed individually into sterile bags, packed on ice, and transported to the laboratory for evaluation. Overall, *Campylobacter* was found in 7 of 55 immature follicles, 12 of 47 mature follicles, and 41 of 55 ceca.

Figure 4.4   Breeder hen ovarian follicle. (Image courtesy of Jeanna Wilson.)

Campylobacter was found in at least one of each sample of mature follicles and in the ceca of hens in each of the five trials. *Salmonella* was found in 0 of 55 immature follicles, 1 of 47 mature follicles, and 8 of 55 ceca. The authors concluded that the recovery rate of *Salmonella* from late-life broiler breeder hen ovarian follicles was relatively low; however, the recovery rate of *Campylobacter* from the hen ovarian follicles was reasonably high, suggesting that these breeder hens could be infecting fertile hatching eggs.

Studies by Vizzier-Thaxton et al. (2006) to determine whether rooster semen (Figure 4.5) is a possible source of transmission to hens for colonization were conducted; they evaluated the association of both *Salmonella* and *Campylobacter* spp. to segments (head, midpiece, and tail) of individual spermatozoa after artificial inoculation. *Salmonella typhimurium*, *Salmonella heidelberg*, and *Salmonella montevideo* or *Campylobacter jejuni* were added to a freshly collected (by abdominal massage) aliquot of pooled semen from roosters housed in individual cages (Vizzier-Thaxton et al., 2006). The semen and bacteria solutions were incubated for 1 hour at room temperature. Individual samples were then subjected to both scanning (JSM-5800) and transmission (JEM-1210) electron microscopy. The scanning electron microscopy showed that *Salmonella* was associated with all three segments (head, midpiece, and tail) of the rooster spermatozoa and apparently equally distributed among those segments. *Campylobacter* was mainly associated with the midpiece and tail segments; few isolates were located on the head segment. Transmission electron microscopy showed apparent attachment of *Salmonella* and *Campylobacter*

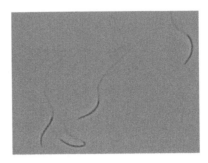

Figure 4.5 Rooster spermatozoa. (Used with permission from John Parrish. http://www.ansci.wisc.edu/jjp1/ansci_repro/lab/lab9/sperm_images/sperm_ images.html.)

to the spermatozoa. These studies by Cox et al. (2005) and Vizzier-Thaxton et al. (2006) showed that it is possible for the rooster to inseminate and inoculate the breeder hens during mating.

Internal contamination of eggs by *Salmonella* Enteritidis has been a significant source of human illness for several decades and is the focus of a recently proposed U.S. Food and Drug Administration regulatory plan (Gast et al., 2007). *Salmonella heidelberg* has also been identified as an egg-transmitted human pathogen. The deposition of *Salmonella* inside eggs is apparently a consequence of reproductive tissue colonization in infected laying hens, but the relationship between colonization of specific regions of the reproductive tract and deposition in different locations within eggs is not well documented. Gast et al. (2007) conducted studies in which groups of laying hens were experimentally infected with large oral doses of *Salmonella heidelberg*, *Salmonella* Enteritidis phage type 13a, or *Salmonella* Enteritidis phage type 14b. For all of these isolates, the overall frequency of ovarian colonization (34.0%) was significantly higher than the frequency of recovery from either the upper (22.9%) or lower (18.1%) regions of the oviduct. The authors found no significant differences between the frequencies of *Salmonella* isolation from egg yolk or egg albumen (4.0% and 3.3%, respectively). Significant differences were observed between *Salmonella* isolates with regard to the frequency of recovery from eggs, but not in the frequency or patterns of recovery from reproductive organs. Gast et al. (2007) reported that, accordingly, although the ability of these *Salmonella* isolates to colonize different regions of the reproductive tract in laying hens was reflected in deposition in both yolk and albumen, there was no indication that any specific affinity of individual isolates for particular regions of this tract produced distinctive patterns of deposition in eggs.

The relationship between colonization of broiler breeder hens with *Salmonella* and vertical transmission of *Salmonella* through the hatchery and to broilers has been firmly established in the research literature. Programs are being implemented worldwide to break this cycle. The following chapter reviews some of these approaches.

References

Bailey J S, Cox N A, Craven S E, and Cosby D E (2002), Serotype tracking of *Salmonella* through integrated broiler chicken operations, *Journal of Food Protection*, 65, 742–745.

Bailey J S, Stern N J, Fedorka-Cray P, Craven S E, Cox N A, Cosby D E, Ladely S, and Musgrove M T (2001), Sources and movement of *Salmonella* through integrated poultry operations: A multistate epidemiological investigation, *Journal of Food Protection*, 64, 1690–1697.

Barnes E M, Mead G C, Barnum D A, and Harry E G (1972), The intestinal flora of the chicken in period 2 to 6 weeks of age with particular reference to the anaerobic bacteria, *British Poultry Science*, 13, 311–326.

Baskerville A, Humphrey T J, Fitzgeorge R B, Cook R W, Chart H, Rowe B, and Whitehead A (1992), Airborne infection of laying hens with *Salmonella* enteritidis phage type 4, *Veterinary Record,* 130, 395–398.

Berrang M E, Frank J F, Buhr R J, Bailey J S, Cox N A, and Mauldin J M (1998), Eggshell characteristics and penetration by *Salmonella* through the productive life of a broiler breeder flock, *Poultry Science*, 77, 1446–1450.

Bryan F L, and Doyle M P (1995), Health risks and consequences of *Salmonella* and *Campylobacter jejuni* in raw poultry, *Journal of Food Protection*, 58, 326–344.

Cox N A, Bailey J S, Berrang M E, Mauldin J M, and Blankenship L C (1990), Presence and impact of salmonellae contamination in the commercial integrated broiler hatchery, *Poultry Science*, 69, 1606–1609.

Cox N A, Bailey J S, Mauldin J M, Blankenship L C, and Wilson J L (1991), Research note: Extent of salmonellae contamination in breeder hatcheries, *Poultry Science*, 70, 416–418.

Cox N A, Bailey J S, Richardson L J, Buhr R J, Cosby D E, Wilson J L, Hiett K L, Siragusa G R, and Bourassa D V (2005), Presence of naturally occurring *Campylobacter* and *Salmonella* in the mature and immature ovarian follicles of late-life broiler breeder hens, *Avian Diseases*, 49, 285–287.

Cox N A, Berrang M E, and Cason J A (2000), *Salmonella* penetration of egg shells and proliferation in broiler hatching eggs—A review, *Poultry Science*, 79, 1571–1574.

Davies R H, and Wray C (1996), Persistence of *Salmonella* Enteritidis in poultry units and poultry food, *British Poultry Science*, 37, 589–596.

Gast R K (1993), Detection of *Salmonella* enteritidis in experimentally infected laying hens by culturing pools of egg contents, *Poultry Science*, 72, 267–274.

Gast R K, and Beard C W (1990), Production of *Salmonella* enteritidis contaminated eggs by experimentally infected hens, *Avian Diseases*, 34, 438–446.

Gast R K, and Beard C W (1992), Detection and enumeration of *Salmonella* Enteritidis in fresh and stored eggs laid by experimentally infected hens, *Journal of Food Protection*, 55, 152–156.

Gast R K, Holt P S, Moore R W, Guraya R, and Guard-Bouldin J (2007), Colonization of specific regions of the reproductive tract and deposition at different locations inside eggs laid by hens infected with *Salmonella* Enteritidis or *Salmonella heidelberg*, *Avian Diseases*, 51, 40–44.

Gregory D W (1948), *Salmonella* infections of turkey eggs, *Poultry Science*, 27, 359–366.

Henken A M, Frankena K, Goelema J O, Graat E A M, and Noordhuizen J P T M (1992), Multivariate epidemiological approach to salmonellosis in broiler breeder flocks, *Poultry Science*, 71, 838–843.

Hoover N J, Kenney P B, Amick J D, and Hypes W A (1997), Preharvest sources of *Salmonella* colonization in turkey production, *Poultry Science*, 76, 1232–1238.

Humphrey T J, Baskerville A, Chart H, Rowe B, and Whitehead A (1992), Infection of laying hens with *Salmonella* enteritidis PT4 by conjunctival challenge, *Veterinary Record*, 131, 386–388.

Humphrey T J, Baskerville A, Mawer S, Rowe B, and Hopper S (1989), *Salmonella* enteritidis phage type 4 from the contents of intact eggs: A study involving naturally infected eggs, *Epidemiology and Infection*, 103, 415–423.

Humphrey T J, Whitehead A, Gawler A H L, Henley A, and Rowe B (1991), Numbers of *Salmonella* Enteritidis in the contents of naturally contaminated hens eggs, *Epidemiology and Infection*, 106, 489–496.

Jones F R, Axtell C, Tarver F R, Rives D V, Scheideler S E, and Wineland M J (1991), Environmental factors contributing to *Salmonella* colonization of chickens, in *Colonization Control of Human Bacterial Enteropathogens in Poultry*, ed L C Blankenship, Academic Press, San Diego, CA, 3–20.

Keller L H, Benson C E, Krotec K, and Eckroade R J (1995), *Salmonella* Enteritidis colonization of the reproductive tract and forming and freshly laid eggs of chickens, *Infection and Immunity*, 63, 2443–2449.

Milner K C, and Shaffer M F (1952), Bacteriologic studies of experimental *Salmonella* infections in chicks, *Journal of Infectious Diseases*, 90, 81–96.

Miyamoto T, Baba E, Tanaka T, Sasai K, Fukata T, and Arakawa A (1997), *Salmonella* enteritidis contamination of eggs from hens inoculated by vaginal, cloacal and intravenous routes, *Avian Diseases*, 41, 296–303.

Poppe C, Johnson R P, Forsberg C M, and Irwin R J (1992), *Salmonella* enteritidis and other *Salmonella* in laying hens and eggs from flocks with *Salmonella* in their environment, *Canadian Journal of Veterinary Research*, 56, 226–232.

Reiber M A, McInroy J A, and Conner D E (1995), Enumeration and identification of bacteria in chicken semen, *Poultry Science*, 74, 795–799.

Sander J, Hudson C R, Dufour-Zavala L, Waltman W D, Lobsinger C, Thayer S G, Otalora R, and Maurer J J (2001), Dynamics of *Salmonella* contamination in a commercial quail operation, *Avian Disease*, 45, 1044–1049.

Shivaprasad H L, Timoney J F, Morales S, Lucio B, and Baker R C (1990), Pathogenesis of *Salmonella* Enteritidis infection in laying chickens. I. Studies on egg transmission, clinical signs, fecal shedding, and serologic responses, *Avian Diseases*, 34, 548–557.

Timoney J F, Shivaprasad H L, Baker R C, and Rowe B (1989), Egg transmission after infection of hens with *Salmonella* Enteritidis phage type 4, *Veterinary Record*, 125, 600–601.

Vizzier-Thaxton Y, Cox N A, Richardson L J, Buhr R J, McDaniel C D, Cosby D E, Wilson J L, Bourassa D V, and Ard M B (2006), Apparent attachment of *Campylobacter* and *Salmonella* to broiler breeder rooster spermatozoa, *Poultry Science*, 85, 619–624.

Chapter 5

Salmonella Intervention in Breeders

5.1 Introduction

Numerous studies have demonstrated that *Salmonella* is passed from breeder chickens to their progeny via production of infected eggs (internally and externally). Thus, intervention at the level of the breeder chicken is essential for controlling this bacterium.

5.2 Effect of Disinfection of Hatching Eggs and Sanitization of Farms

As early as 1992, Henken and colleagues studied data from 111 broiler breeder flocks over a 5-year period. The analysis revealed that farms with egg disinfection tubs, good biosecurity, and a large feed mill had 46.1 times less risk of being *Salmonella* positive than flocks housed at farms that did not have these interventions or that were supplied feed from a small feed mill. However, many breeders refuse to disinfect commercial fertile hatching eggs because of a fear of increasing *Salmonella* penetration across the eggshell.

Fris and Van den Bos (1995) reported that the Dutch poultry industry has attempted to prevent vertical transmission of *Salmonella* from breeders to broilers using a top-down approach with particular emphasis on controlling *Salmonella* serotype Enteritidis (SE). However, the efficacy of this program is now affected by the increasing role of horizontal transmission, with biosecurity on the poultry farm gaining importance (Fris and Van den Bos, 1995). To assess the actual level of preventive hygiene and to identify risk factors involved, an inquiry was held among a representative number of broiler breeder farms. From these inquiries, it became evident that the hygiene conditions on Dutch broiler breeder farms can and must be improved (Fris and Van den Bos, 1995). From a matched case-control study carried out among SE-infected and SE-free farms, the occurrence of SE infection could be explained for about 30% by factors concerning preventive hygiene, the surroundings of the farm, and the farm itself. Thus, other factors, not included in the study, may also be important. No particular key factors were found. Preventive hygiene can certainly reduce the infection risk, but only a comprehensive package of measures can do so. Fris and Van den Bos (1995) concluded that the Dutch Product Board for Poultry and Eggs and the Dutch Organization of Poultry Farmers have now agreed to strict biosecurity programs to control horizontal transmission of SE, but these will require considerable technical and financial investments.

Davies and Wray (1996) sampled three broiler breeder houses on three different sites before and after cleaning and disinfection for the presence of *Salmonella*. None of the farms achieved total elimination of SE from the poultry house environment; however, substantial improvements were observed after errors in the cleansing and disinfection protocol in the first house had been corrected. Fundamental errors such as overdilution and inconsistent application of disinfectants were noted, despite supervision of the process by technical advisors. In each of the three poultry units, failure to eliminate a mouse population that was infected with SE was likely to be the most important hazard for the next flock.

5.3 Effect of Medication

In 1997, Reynolds et al. attempted to medicate breeding flocks of domestic fowl (*Gallus gallus*) using an antimicrobial treatment followed by competitive exclusion in 13 trials between February and September 1993. This approach was being used as an alternative to the Swedish model such that positive flocks would not have to be slaughtered but would be treated instead. In each trial, the flock had been confirmed as naturally infected with *Salmonella enterica* serovar Enteritidis, and the effect of treatment was determined on *Salmonella* isolation from internal tissues. Of the 11 trials in which enrofloxacin was used to medicate the flocks, a long-term reduction of *Salmonella* was observed in 2, and a short-term reduction was measured in birds from another 5 trials. SE was isolated from birds after treatment in four other trials with enrofloxacin and in two trials of medication with amoxicillin. The authors concluded that enrofloxacin significantly reduced the prevalence of SE in tissues from birds and reduced the level and prevalence of *Salmonella* in the bird's environment. No *Salmonella* was identified in statutory meconium samples taken from the hatched chicks derived from the flocks after treatment. The program of antibiotic treatment and competitive exclusion offers an alternative to slaughter, but the approach must be part of a coordinated program that will effect a decrease in the prevalence of SE over time by contemporary use of disease security measures (Reynolds et al., 1997). Unfortunately, the U.S. Food and Drug Administration has banned the use of enrofloxacin in poultry operations. The final decision of the FDA commissioner was dated July 27, 2005, and ordered the approval for NADA 140-828 (enrofloxacin) to be withdrawn pursuant to regulation § 512(e)(1)(B) of the Food, Drug, and Cosmetic Act. The order became effective September 12, 2005.

5.4 Effect of Vaccines on Breeder Chickens

Broiler breeder chicks of two different genetic lines were evaluated for early antibody response to SE vaccine (Kaiser et al., 1998). The authors evaluated antibody responses to three separate dosages of SE vaccine administered at 22 days of age at Days 3, 6, and 10 postvaccination. Within each genetic line, antibody levels at 10 days postvaccination were significantly higher than at either 3 or 6 days postvaccination. At all vaccine dosages, there was a significant antibody response difference between the genetic lines at 6 and 10 days postvaccination. In this study, the authors were able to demonstrate that vaccine dosage significantly affected antibody levels in one of the two genetic lines, and that there was a genetic component of early antibody response to SE vaccine in broiler breeder chicks (Kaiser et al., 1998).

From August 1995 until December 1997, Feberwee et al. (2000) studied the effect of adding an SE vaccination to a certified standardized biosecurity program in a field trial in the Netherlands. In this field trial, two groups of broiler breeder flocks with increased infection risk were vaccinated, one group with VAC-T/TALOVAC log SE (Group A) and the second group with SALENVAC (Group B). The determination of increased infection risk in Groups A and B was based on an SE infection history in which flocks were either previously infected and treated (PIT) or had other risk factors than previously infected and treated (OPIT). SE infections in both vaccinated groups were assessed by monitoring according to the Dutch *Salmonella* control program. The authors noted that, under field conditions, designation of a vaccinated and a control group on the farm was not possible; however, in the same time period as the vaccinated groups were being evaluated, 608 nonvaccinated flocks (Group C) were hatched and monitored according to the Dutch *Salmonella* control program. The level of occurrence of SE infection in the flock for the vaccinated groups was compared with the flock-level occurrence of SE infection in the nonvaccinated group based on comparability of infection risk. The proportion of SE-infected flocks with risk factor PIT in the vaccinated groups was not significantly different from that in the nonvaccinated Group C. Feberwee et al. (2000) reported that only the proportion of SE-infected flocks with a risk of reinfection in the vaccinated Group B (0%) was significantly lower ($p = 0.02$) than in the nonvaccinated Group C (18%). The fact that no significant result was found in favor of Group A is because of the small number of flocks in this part of the study. The authors concluded that, based on the conditions of the setup of this trial, it can only be concluded that there is an indication that vaccination contributes in the reduction of SE reinfection in broiler breeder flocks. However, it is clear that in this study, vaccination was not a powerful method of intervention.

Kaiser et al. (2002) stated that the relationship between antibody response to SE vaccine and internal organ carriage of SE is not fully understood. The genetic relationship, therefore, between postchallenge SE burden and antibody response to SE vaccine was determined in broiler breeder chicks (Kaiser et al., 2002). Sibling chicks from a broiler breeder male line were either inoculated with a pathogenic SE strain or vaccinated with a commercial SE vaccine. Spleen, liver, cecal wall, and cecal content samples from 120 SE-challenged chicks were cultured for enumeration of *Salmonella*. Unchallenged chicks ($n = 314$) were vaccinated at 11 days of age, and serum samples were taken at 10 days postvaccination. Antibody responses to vaccination and the number of SE in cecal content cultures were negatively correlated (-0.772), demonstrating that genetic potential for greater antibody response to SE vaccine is associated with lesser SE bacterial burden in the cecal content of broiler breeder chicks. The authors concluded that genetic selection for vaccine antibody responsiveness can lower bacterial burden in the gut lumenal content, thus potentially reducing contamination of poultry products at processing (Kaiser et al., 2002).

5.5 Effect of Probiotics and Maternal Vaccination

The effects of probiotics and maternal vaccination with an inactivated SE vaccine on day-old chicks challenged with SE were evaluated by Avila et al. (2006). Groups of breeder chickens with or without probiotics and groups of breeders nonvaccinated, vaccinated intramuscularly, or vaccinated intraperitoneally were tested. The authors found that the number of SE bacteria per chick and the time interval between housing and introduction of seeder birds (challenge) were 1.6×10^8 and 1 hour (Trial I), 1.8×10^6 and 12 hours (Trial II), and 1.2×10^4 and 24 hours (Trial III). SE recovery was assessed in ceca and liver at 3, 5, and 7 days postchallenge, and the number of colony-forming units (CFU) in the ceca was evaluated at 5 and 7 days postchallenge. The number of SE ($\log_{10}$ CFU) in the ceca was reduced by 0.56 $\log_{10}$ (from 7.59 to 7.03) and 1.45 $\log_{10}$ (from 7.62 to 6.17) because of the treatment with probiotics in Trials II and III, respectively. Avila et al. (2006) concluded that the greater reduction in Trial III indicated the importance of the early use of probiotics on the prevention of SE infection in the birds. Treatment with probiotics resulted in a smaller number of SE-positive livers after 5 days postchallenge on Trial III. Interestingly, the authors did not find any significant effect of maternal vaccination on the number of SE CFU in the ceca; however, a significant effect of maternal vaccination on the SE CFU was observed in the liver at 5 days postchallenge.

In 2006, Schaefer et al. examined the effect of two turkey breeder hen groups at different ages (33 or 55 weeks of age) on performance, intestinal histology, and inflammatory immune response of female turkey poults grown to market weight. At Days 10, 24, and 65 posthatch, turkey poults were vaccinated with lipopolysaccharide (LPS, from ST; 0.5 mg/kg of body weight intra-abdominally) or not vaccinated (control), and intestinal histology and plasma haptoglobin were assessed at 24 hours postadministration. The authors found that, in control birds, intestinal villus length was greater for poults from the older breeder flock ($p < 0.05$), as was crypt depth ($p < 0.05$ for Days 11 and 25). Plasma haptoglobin levels did not change in 11-day-old poults after LPS administration, but they increased with LPS at Days 25 and 66 posthatch ($p < 0.05$ for each). At Day 66 posthatch, poults from the younger flock had increased haptoglobin levels post-LPS compared with those from the older breeder flock ($p < 0.05$). LPS administration increased villus width in the jejunum and ileum ($p < 0.05$ for each), increased lamina propria width in the duodenum and ileum ($p < 0.05$ for each), and decreased ileum crypt depth ($p < 0.05$). Schaefer et al. (2006) concluded that poults from the older breeder flock had reduced inflammatory responses, even at 9 to 10 weeks posthatch, even though performance was similar in poults from the two flocks by this age.

Inoue et al. (2008) stated that young poultry are very susceptible to SE infections because of the absence of complete intestinal flora colonization and an immature immune system. The authors conducted a study to evaluate the role of passive immunity on the resistance of young birds against early infections caused by SE. The progeny of broiler breeders were vaccinated with an oil emulsion bacterin and were compared to the progeny of unvaccinated birds. Efficacy was determined by challenging birds at Days 1 and 14 with SE phage type 4. After challenge at 1 day of age, the progeny of vaccinated birds presented a significantly lower number ($\log_{10}$) of SE ($p < 0.05$) in liver (2.21), spleen (2.31), and cecal contents (2.85) compared with control groups (2.76, 3.02, and 6.03, respectively; Figure 5.1).

Inoue et al. (2008) observed that examination of the internal organs 3 days after infection revealed that 28% of the birds (7/25) from vaccinated breeders were positive, whereas 100% (25/25) of the chicks derived from unvaccinated birds were positive. Moreover, birds that were challenged at 14 days of age showed a lower number of positive samples compared with those challenged at 1 day of age, and the progeny of vaccinated birds presented statistically lower numbers (2.11 vs. 2.94). In this study, age influenced the susceptibility of birds to SE infections: In control groups, the number of positive birds at 14 days of age (9/25) was lower when compared with the group infected at 1 day of age (25/25). The number of positive fecal samples of the progeny of vaccinated birds was significantly lower (36) than those of the control group (108) after challenge at 1 day of age. Passive antibodies were detectable by enzyme-linked immunosorbent assay (ELISA) up to 21 days of age in unchallenged progeny of vaccinated birds (Inoue et al., 2008). For the control group, antibodies were detected by ELISA 14 days after challenge. The authors concluded that these results indicated a significant contribution of breeder

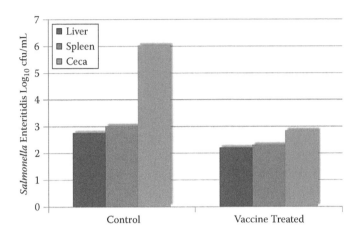

Figure 5.1 *Salmonella* **Enteritidis colonization of the liver, spleen, and ceca of vaccine-treated and untreated control breeder chickens. (From Inoue A Y, Paiva J B, Sterzo E V, Berchieri, A V, Jr, and Bernardino A, 2008, Passive immunity of progeny from broiler breeders vaccinated with oil-emulsion bacterin against** *Salmonella* **Enteritidis,** *Avian Diseases,* **52, 567–571.)**

vaccination by increasing the resistance of the progeny against early SE infections. However, the bacteria were not completely eliminated, suggesting that additional procedures are needed to effectively control SE infections (Inoue et al., 2008).

Most efforts to control *Salmonella* in breeder flocks in the United States have concentrated on vaccination programs. In some cases, they are very effective; however, in others they do not have any effect. In working with companies that were vaccinating 100% of their breeders, plants receiving the broilers from these flocks have reported a 100% *Salmonella*-positive rate. These data indicate that the vaccination program in these operations is having no impact (S. M. Russell, unpublished data, 2004).

5.6 Effect of Reducing *Salmonella* in the Air of Layers

Airborne dust in poultry housing is known to be one of the primary means by which disease-causing organisms are spread throughout a house (Mitchell et al., 2004). An electrostatic space charge system (ESCS) was used to reduce airborne dust in a small-scale broiler breeder house. The system used ceiling fans to distribute negatively charged air throughout the room and to move negatively charged dust downward toward the grounded litter, where most of it would be captured. Mitchell et al. (2004) found that this system significantly ($p < 0.0001$) reduced airborne dust by an average of 61%, ammonia by an average of 56% ($p < 0.0001$), and airborne bacteria by 67% ($p < 0.0001$). The ESCS was shown to be a reliable and easily maintained system for reducing airborne dust, ammonia, and bacteria in a small broiler breeder house. Results of this study combined with the results of related ESCS studies suggest that the system could probably be scaled up to full-size production houses for poultry or other animals for dust reduction, pathogen reduction, and possibly ammonia reduction (Mitchell et al., 2004).

5.7 Control of *Salmonella* in Breeders Using Elimination of Positive Flocks

As early as 1990, specific countries within Europe began aggressive programs to eliminate *Salmonella* in breeder and layer chickens. Sweden, Denmark, and the Netherlands have enacted programs to dramatically reduce the prevalence on *Salmonella* in breeder chickens and subsequently in broiler chickens. Governmental regulations in Sweden, dating to 1961, were introduced as a result of a large *Salmonella* epidemic in 1953. Since then, Sweden maintains an active, organized system for controlling *Salmonella* in poultry (Wierup et al., 1992). The objective of this system is to deliver *Salmonella*-free poultry to consumers. Chickens delivered to the plant for slaughter must be free from *Salmonella* by applying the following strategies: (1) prevent contamination of all parts of the production chain, (2) monitor the production chain at critical control points to detect if *Salmonella* contamination has occurred, and (3) undertake actions necessary to fulfill the objective of the control when *Salmonella* contamination is detected (Wierup et al., 1992).

In 1990, Sweden implemented an aggressive program to eliminate *Salmonella* in all layer chickens. Approximately 90% of the layer flocks were voluntarily tested by bacteriological examination of pooled fecal samples for *Salmonella* before slaughter. If SE was isolated, the entire breeder flock was destroyed. This voluntary *Salmonella* control program was extended to all breeder chickens and hatcheries in Sweden in 1991. The control program has been directed at all serotypes of *Salmonella*, and any breed of grandparent chickens that are imported must also be evaluated. The authors reported that this testing program is the main reason why Sweden has not been involved in the worldwide spread of different phage types of SE (Wierup et al., 1992).

To assist in preventing transmission of *Salmonella* from the breeder chickens to the egg, then to broilers in Sweden, a variety of measures are used. All feed for layer chickens during the rearing period must be heat treated, and now, the use of heat-treated feed is becoming gradually more common during grow out. Wierup et al. (1992) reported that, because of these and other measures, SE has not been isolated from broilers in Sweden since 1972.

In 1994, Edel reported that until 1988, the percentage of SE in poultry was stable, and was less than 4% of the total number of strains isolated from humans in the Netherlands, but rose significantly during 1988 to 8%. There was also an increase in the number of times SE was isolated from poultry (Edel, 1994). Early in 1988, the Produktschap voor Pluimvee en Eieren (PPE), translated as Commodity Board for Poultry and Eggs, was confronted by the fact that SE had been found in a grandparent flock (broiler breeders). This flock was originally from eggs imported from the United Kingdom. Suddenly, an important economic problem occurred in broiler flocks, because chick mortality rose, in some cases to 10%, and the surviving birds did not perform well. At the same time, an SE problem also occurred in a layer parent flock that was hatched from eggs imported from Germany. These events led the Netherlands to institute an SE eradication program in poultry breeder flocks. The country performed a top-down approach, which meant that by eliminating SE from the top end (breeding stock) and utilizing good hygiene practices throughout the industry, the organism should progressively be cleared from the whole national stock (Edel, 1994). Thus, each year, all poultry breeder flocks (approximately 2,300) were screened for the presence of SE by bacteriological examination of cecal droppings until April 1 of 1992. After that date, screening was carried out with a double antibody sandwich blocking (DAS blocking) ELISA method (Edel, 1994). The industry found that treatment of SE-positive flocks, even while in production, with Baytril® (enrofloxacin) in combination with using competitive exclusion flora eliminated SE just as well as slaughtering these flocks. The top-down approach used to eliminate SE in poultry seems to have greatly affected SE infection in humans because, in the Netherlands, no further increase in the number of human infections was observed from 1991 to 1994.

As in Sweden and the Netherlands, Denmark employed a similar program. Wegener et al. (2003) reported that Sweden remains essentially free from the *Salmonella* problems typical for most other industrialized countries. Unfortunately, other countries cannot apply the Swedish model of *Salmonella* control, which requires near freedom from *Salmonella* in domestic food animal production from the onset (Wegener et al., 2003). In the European Union, the Zoonosis Directive was an attempt to initiate a European Union-wide control effort against food-borne zoonoses, particularly *Salmonella* in broiler chickens and layer hens; however, most E.U. countries found that they either could not or would not implement the directive, which did not permit use of vaccines, antimicrobial drugs, or both as elements in the control program of *Salmonella* in broiler chickens or layer hens (Wegener et al., 2003).

Rapid increases in the incidence in human salmonellosis in the second half of the 1980s in Denmark was attributed to the spread of *Salmonella* in broiler chickens. Because of this link to poultry, a targeted national control program was initiated. Initially, the aim of the program was to reduce *Salmonella* in broiler flocks to less than 5% prevalence (Wegener et al., 2003). The program was developed based on the concept of eliminating *Salmonella* from the breeders, which would then theoretically ensure that broilers and processed products would be free from *Salmonella*. Wegener et al. (2003) reported that infected flocks of breeder chickens were being destroyed, and infected birds were being slaughtered. The intensive testing program developed gradually over time. Birds from *Salmonella*-positive flocks

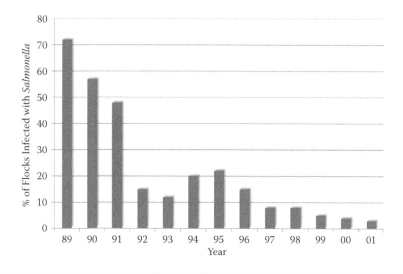

Figure 5.2 Percentage of flocks infected with *Salmonella* in Denmark from 1989 to 2001 evaluated 2–3 weeks prior to slaughter (*N* > 4,000 flocks/year). (From Wegener H C, Hald T, Wong D L F, Madsen M, Korsgaard H, Bager F, Gerner-Smidt P, and Mølbak K, 2003, Salmonella control programs in Denmark, *Emerging Infectious Diseases*, 9, 774–780. http://www.cdc.gov/ncidod/EID/vol9no7/03-0024.htm.)

are slaughtered on separate slaughter lines or late in the day to avoid cross contamination of *Salmonella*-negative birds. One incentive the Danes give poultry growers is that farmers get a better price for birds from *Salmonella*-free flocks, and slaughterhouses can use the label "*Salmonella* free" for birds that meet criteria determined by the authorities (Wegener et al., 2003). The effect of this program may be seen in Figure 5.2.

The human salmonellosis prevalence associated with the three major sources of human salmonellosis in Denmark from 1988 to 2001 was reduced significantly as a result of implementation of this program in broilers, layers, and pigs. The broiler-associated salmonellosis prevalence (cases/100,000) has been reduced by greater than 95%, from 30.8 in 1988 to 0.5 in 2001; the egg-associated salmonellosis incidence has been reduced by nearly 75%, from 57.7 in 1997 to 15.5 in 2001; and the pork-associated salmonellosis incidence has been reduced by greater than 85%, from 22.0 in 1993 to 3.0 in 2001 (Wegener et al., 2003).

5.8 Cost of Eliminating Positive Flocks

The cost of the program in Denmark to eliminate positive flocks is very high; however, it was calculated to be well worth the expense. Assuming that salmonellosis associated with each of the major sources would have remained at the precontrol program prevalence (i.e., if no action had been taken to curb the problem), Wegener et al. (2003) calculated a hypothetical "no control" salmonellosis prevalence. The societal costs, in the absence of the existing control programs, would thus have been $41 million per year. Thus, it was calculated that, in 2001, Denmark saved $25.5 million by controlling *Salmonella*. The estimated annual *Salmonella* control costs from 2000 onward have been calculated as approximately $14.1 million. What is most critical to producers in the United States considering this approach is that the cost of this program is borne almost exclusively by the animal producers and the food industry, which suggests that the costs are passed on to consumers through higher food prices. The average price for a small frozen chicken in Denmark is approximately $3.82 and for a large free-range chicken carcass is up to $15.28. The Danish people paid $4.36 for a dozen medium eggs in February 2008 (http://www.visitdenmark.com/uk/en-gb/menu/turist/turistinformation/fakta-az/madpriser.htm). The cost to the broiler producer has been calculated to be $0.02/kg of broiler meat or egg (Wegener et al., 2003). Thus, the cost of such control programs in terms of increasing food prices is significant. In contrast, in the United States, in January 2011, the average cost of a dozen eggs was $1.19, with eggs in Denmark costing 3.66 times as much as eggs in the United States (http://www.thepeoplehistory.com/pricebasket.html).

References

Avila L A F, Nascimento V P, Salle C T P, and Moraes H L S (2006), Effects of probiotics and maternal vaccination on *Salmonella* Enteritidis infection in broiler chicks, *Avian Diseases*, 50, 608–612.

Davies R H, and Wray C (1996), Studies of contamination of three broiler breeder houses with *Salmonella* enteritidis before and after cleansing and disinfection, *Avian Diseases*, 40, 626–633.

Edel W (1994), *Salmonella* Enteritidis eradication programme in poultry breeder flocks in the Netherlands, *International Journal of Food Microbiology*, 21, 171–178.

Feberwee A, De Vries T S, Elbers A R W, and De Jong W A (2000), Results of a *Salmonella* Enteritidis vaccination field trial in broiler-breeder flocks in the Netherlands, *Avian Diseases*, 44, 249–255.

Fris C, and Van Den Bos J (1995), A retrospective case-control study of risk factors associated with *Salmonella* enterica subsp. enterica serovar Enteritidis infections on Dutch broiler breeder farms, *Avian Pathology*, 24, 255–272.

Henken A M, Frankena K, Goelema J O, Graat E A M, and Noordhuizen J P T M (1992), Multivariate epidemiological approach to salmonellosis in broiler breeder flocks, *Poultry Science*, 71, 838–843.

Inoue A Y, Paiva J B, Sterzo E V, Berchieri A V, Jr, and Bernardino A (2008), Passive immunity of progeny from broiler breeders vaccinated with oil-emulsion bacterin against *Salmonella* Enteritidis, *Avian Diseases*, 52, 567–571.

Kaiser M G, Lakshmanan N, Wing T, and Lamont S J (2002), *Salmonella* enterica serovar Enteritidis burden in broiler breeder chicks genetically associated with vaccine antibody response, *Avian Diseases*, 46, 25–31.

Kaiser M G, Wing T, and Lamont S J (1998), Effect of genetics, vaccine dosage, and postvaccination sampling interval on early antibody response to *Salmonella* Enteritidis vaccine in broiler breeder chicks, *Poultry Science*, 77, 271–275.

Mitchell B W, Richardson L J, Wilson J L, and Hofacre C L (2004), Application of an electrostatic space charge system for dust, ammonia, and pathogen reduction in a broiler breeder house, *Applied Engineering in Agriculture*, 20, 87–93.

Reynolds D J, Davies R H, Richards M, and Wray C (1997), Evaluation of combined antibiotic and competitive exclusion treatment in broiler breeder flocks infected with *Salmonella* enterica serovar Enteritidis, *Avian Pathology*, 26, 83–95.

Schaefer C M, Corsiglia C M, Mireles A, Jr, and Koutsos E A (2006), Turkey breeder hen age affects growth and systemic and intestinal inflammatory responses in female poults examined at different ages posthatch, *Poultry Science*, 85, 1755–1763.

Wegener H C, Hald T, Wong D L F, Madsen M, Korsgaard H, Bager F, Gerner-Smidt P, and Mølbak K (2003), *Salmonella* control programs in Denmark, *Emerging Infectious Diseases*, 9, 774–80. http://www.cdc.gov/ncidod/EID/vol9no7/03–0024.htm.

Wierup M, Wahlstrom H, and Engstrom B (1992), Experience of a 10-year use of competitive exclusion treatment as part of the *Salmonella* control programme in Sweden, *International Journal of Food Microbiology*, 15, 287–291.

Chapter 6

The Role of the Hatchery in *Salmonella* Transfer

6.1 Introduction

Many opportunities exist for *Salmonella* to be transferred from contaminated eggs to uninfected baby chicks during the hatching process. Cox et al. (2000) published an excellent review of this subject. *Salmonella* may be found in the nest boxes where breeders lay eggs, in the cold storage egg room at the breeder farm, on the truck that transports baby chicks to the grow-out houses, or in the hatchery environment. All of these situations may cause horizontal contamination of the eggs with *Salmonella*. Once transferred, the *Salmonella* is carried on the surface of the shell or just beneath the shell if it is able to penetrate the shell. One mechanism for natural contamination of the eggs is that, when moist, freshly laid eggs are cooled from the body temperature of the hen to the air temperature, and the internal contents of the egg shrink, which pulls the bacteria into the shell through pores (Cox et al., 2000).

6.2 How Fertilized Eggs Become Contaminated with *Salmonella*

The breeder hen's nest becomes contaminated with *Salmonella* when the bird brings soil and feces into it. Smeltzer et al. (1979) demonstrated that eggs laid in wet, dirty nests or on the floor are more likely to be contaminated with bacteria. It has been known for over 100 years that salmonellae are able to penetrate eggshells. Moreover, numerous studies have shown the ability of *Salmonella* to penetrate and multiply within the contents of both chicken and turkey hatching eggs (Cox et al., 2000). There exists significant variability in terms of how well *Salmonella* can penetrate eggs (Stokes et al., 1956; Humphrey et al., 1989, 1991). Shell attributes (Sauter and Petersen, 1974), pH (Sauter et al., 1977), number of pores on an eggshell (Walden et al., 1956), temperature (Graves and Maclaury, 1962), humidity (Gregory, 1948), and vapor pressure (Graves and Maclaury, 1962) are all factors that may affect penetration (Cox et al., 2000). In spite of the protective effect of the inner and outer eggshell membranes (Baker, 1974), several researchers have demonstrated that *Salmonella* and other bacteria may penetrate these membranes rapidly. Studies have shown that bacteria were able to penetrate 25–60% of inner membranes and 10–15% of albumen in eggs that were inoculated (Muira et al., 1964; Humphrey et al., 1989, 1991). Other scientists observed that penetration of the cuticle and shell of eggs by salmonellae occurred almost immediately in some eggs. In fact, in one egg, penetration below both membranes was detected as early as 6 minutes following shell exposure (Williams et al., 1968). The authors found that once bacteria got past the membranes of hatching eggs, there was no way to prevent their further invasion of the egg contents or developing embryo.

6.3 Fertilized Hatching Egg Sanitation

Davies and Wray (1994) reported that sanitizing eggs is the first barrier in preventing introduction of *Salmonella* contamination into the hatchery premises via the egg surface. The authors stated that sanitizing eggs using disinfectant fogging in the hatchery was insufficient, and that eggs had to be treated using formaldehyde vapors or further egg sanitization through a well-regulated washing machine (Figures 6.1 and 6.2) (Davies and Wray, 1994). Control of *Salmonella* contamination in setters is important because the warm, humid air is able to disseminate *Salmonella* over the surfaces of several batches of eggs within the same incubator (Figures 6.3 and 6.4).

The authors (Davies and Wray, 1994) reported that after 18 days of turning eggs in the setter incubators, the eggs are transferred to hatcher incubators for the final 3 days of incubation. The transfer from egg trays to hatching baskets is often semiautomatic, using multiple-head suction cup machines that transfer a whole tray of eggs in one operation. This speeds up the transfer process, but the suction heads are often contaminated with salmonellae and can cause cross contamination of batches of eggs. Frequent disinfection of surfaces and suction cups of egg transfer machines has resulted in improvements in the *Salmonella* contamination rate of egg transfer equipment in one hatchery (Davies and Wray, 1994). Cox et al. (2000) also noted that the warm temperature of incubation enhances the multiplication of salmonellae (Cox et al., 2000). Rizk et al. (1966b) indicated that salmonellae that

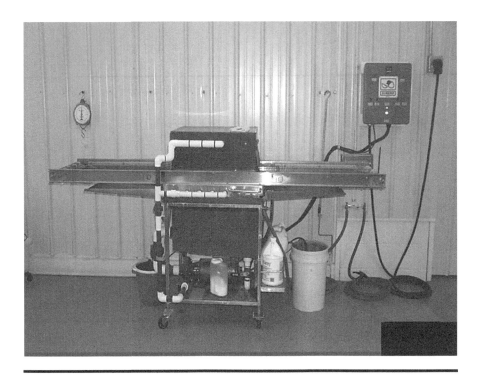

Figure 6.1   Egg-washing machine.

Figure 6.2   Eggs after washing.

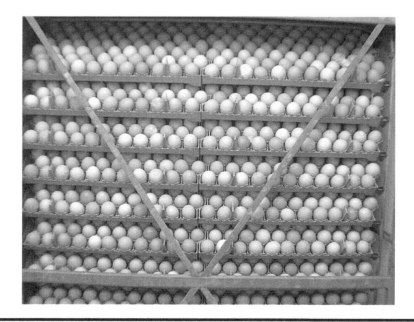

Figure 6.3 Eggs in the setter.

Figure 6.4 Eggs in the setter (different view).

are able to penetrate the eggshell will multiply greatly during incubation because of the high temperature.

6.4 Effect of Contaminated Eggshells

Bailey et al. (1992) reported that contaminated eggshells lead to spread of salmonellae in the hatchery. In 1990 and 1991, Cox et al. reported that breeder and broiler hatcheries were highly contaminated with salmonellae. In the broiler hatchery, *Salmonella* was detected on 71% of eggshell fragments, 80% of chick conveyor belt swab samples, and 74% of samples of pads placed under newly hatched chicks to gather feces. According to Cox et al. (2000), because many eggs coming from a naturally infected flock must be tested to find one that is salmonellae positive, one must wonder how such an increase in contamination occurs.

In 1994, Cason et al. reported that eggs inoculated with high numbers of *Salmonella* Typhimurium still hatch because paratyphoid salmonellae generally do not present a health problem to the chick. In their experiments, 120 unincubated, fertile hatching eggs were inoculated by immersion for 15 minutes in a 16°C physiological saline solution containing 1×10 colony forming units (CFU) per milliliter of a nalidixic-acid-resistant strain of *Salmonella* Typhimurium. When inoculated eggs were transferred to hatchers after 17 to 18 days of incubation, control eggs at the same stage of incubation were added to the same tray and to trays above and below the tray containing the inoculated eggs. Fertile inoculated eggs hatched at a rate of 86%, despite the high level of *Salmonella* contamination, indicating that chicks in eggs contaminated with salmonellae are likely to hatch and may contaminate other chicks in the same hatcher cabinet. Air samples showed a sharp increase in contamination in the hatcher at 20 days of incubation. Approximately 58% of mouth swabs and 90% of chick rinses were *Salmonella* positive in both inoculated and control eggs. In samples from inoculated eggs, *Salmonella* was detected in the digestive tract of 8% of embryos at transfer from incubator to hatcher and in 55% of chicks at hatch. From control eggs, 44% of digestive tracts of hatched chicks were positive, indicating that *Salmonella* in a contaminated hatcher can reach the gut of chicks hatching from *Salmonella*-free eggs before they are removed from the hatcher (Cason et al., 1994). After hatching, *Salmonella* cells are widely disseminated through the hatching cabinet due to rapid air movement by fans to keep the temperature and humidity of the hatching cabinet constant. In another study, Cason et al. (1993) showed that eggs carrying salmonellae on the exterior or in shell membranes could lead to contamination of the chick when the chick pips the eggshell. When the embryos were sampled prior to eggshell pipping, no *Salmonella* were detected on body rinses, but after the shell had been breached, 15% of the chickens were externally contaminated, and 8% had contaminated yolk sacs (Cason et al., 1993). These studies demonstrated the importance of disinfecting eggs during the setting and hatching stages to prevent cross contamination of *Salmonella* to baby chicks during the hatching process.

6.5 Disinfection during Setting and Hatching

The concept of sanitizing eggs using disinfectant has been around for a long time (Pritsker, 1941), and a variety of chemicals and application methods have been evaluated over the years. Unfortunately, much of the information gained in earlier studies may not be applicable because methods to recover sublethally injured cells of *Salmonella* were not used. In addition, when a chemical was shown to be effective, too much time had often transpired prior to chemical treatment. For example, 30 minutes (Bierer et al., 1961; Rizk et al., 1966a), 60 minutes (Frank and Wright, 1956), 3 hours (Gordon et al., 1956), 2 days (Mellor and Banwart, 1965), and 1 to 5 days (Lancaster and Crabb, 1953) were the time intervals between inoculation and application of treatment. Williams and Dillard (1973) demonstrated that the time intervals used in these previous studies were too long to expect the chemical to be effective.

In 1989, the U.S. poultry industry processed 7.5 billion hatching eggs through incubating facilities (Brown, 1989), and that number grew to 9 billion by 2000 (Cox et al., 2000). Because bacteria that invade an egg do not cause extensive decomposition, the chick hatches from the contaminated egg (Maclaury and Moran, 1959). This scenario allows establishment of extensive bacterial reservoirs in commercial hatcheries (Cox et al., 2000). In 1959, Wright et al. tested over 1,000 samples from 120 commercial hatcheries and discovered high microbial populations. More recently, commercial hatcheries have been using a variety of chemical sanitizers and application methods to disinfect the hatchery or eggs, with the belief that the answer lies in either correct choice of chemical or how often or how much to apply. Unfortunately, chemicals seem to have had little effect on salmonellae contamination of hatching eggs, as evidenced by the fact that many researchers have demonstrated the presence of salmonellae in commercial hatcheries (Cox et al., 2000). Goren et al. (1988) isolated salmonellae from three commercial hatcheries in Europe and demonstrated that the *Salmonella* serotypes originating from the hatchery could later be found on processed broiler carcasses. Cox et al. (1990) were able to isolate salmonellae from over 75% of the samples collected from several commercial hatcheries.

Salmonellae have the ability to persist for long periods of time in commercial hatcheries (Cox et al., 2000). Muira et al. (1964) found that over 50% of chick fluff samples were contaminated with salmonellae. When these authors stored hatchery fluff for 4 years at room temperature, 1,000 to 1 million viable *Salmonella* per gram were able to be recovered. Cox et al. (2000) stated that salmonellae-free chickens should be grown and delivered to the processing plant. To accomplish such a lofty goal, the authors stated that salmonellae contamination in fertile hatching eggs and in the hatchery should be controlled. To have any chance of controlling this contamination, hatching egg-sanitizing programs must be applied at the breeder farm level as soon after the egg is laid as is practically possible (Cox et al., 2000).

6.6 Contamination of Baby Chicks

Jones et al. (1991a) and Lahellec and Colin (1985) respectively reported that 5% and 9% of day-old baby chicks were contaminated with *Salmonella*. Jones et al. (1991b) reported that *Salmonella* was found in 9.4% of the yolk sac samples collected from day-old chicks in hatcheries. However, when studying serotypes of *Salmonella* to determine whether there was a link between *Salmonella* found on final product and serotypes of *Salmonella* on chicks in the hatchery, Lahellec and Colin (1985) stated that the serotypes of *Salmonella* found in the hatchery were less important in terms of the ones found on the final product; however, the grow-out environment was the most important source of *Salmonella* serotypes found at slaughter. Other researchers supported this contention when they reported that the environment of the chicken during grow out was the primary source of *Salmonella* contamination (Bailey et al., 1991; Blankenship et al., 1993). Interestingly, Bailey et al. (2002) found that there was poor correlation between the serotypes of *Salmonella* found on the breeder farms and those found in the hatchery. The authors stated that this finding and the fact that similar serotypes were found in the hatchery in both trials suggest that there was an endemic population of *Salmonella* in the hatchery. An association between the serotypes found in the hatchery and those found on the final processed carcasses was observed in both trials. This study confirmed that a successful intervention program for broiler production operations must be multi-faceted, with one component being disinfection in the hatchery. Based on tracking *Salmonella* serotypes, it appears from these studies that the grow-out environment is a greater contributor to contamination than the hatchery.

Corry et al. (2002) conducted a comprehensive investigation of salmonellas in two U.K. poultry companies to find the origins and mechanisms of *Salmonella* contamination. Serovars of *Salmonella* detected during grow out were usually also found in a small proportion of birds on the day of slaughter and on the carcasses at various points during processing (Corry et al., 2002). Little evidence was found indicating that *Salmonella* was spread to large numbers of carcasses during processing. Corry et al. (2002) reported that many of the *Salmonella* serovars found in the feed mills or hatcheries were also detected in the birds during rearing or slaughter. Transportation coops remained contaminated with *Salmonella*, even after washing and disinfection. The authors found that a small number of *Salmonella* serovars predominated in the processing plants of each company, and that these serovars originated from the feed mills. The reasons for transport coop contamination were listed as (1) inadequate cleaning, resulting in residual fecal soiling; (2) disinfectant concentration and temperature was too low; (3) contaminated recycled flume water was used to soak the coops. The conclusion of this study was that efforts to control *Salmonella* infection in broilers need to concentrate on coop cleaning and disinfection and sanitation in the feed mills (Corry et al., 2002).

Bailey et al. (1994) stated that hatchery contamination can result in exposure of newly hatched chicks to salmonellae at a time when they are most susceptible to colonization of their intestinal tract. Eggshell fragments, external rinses on the chicks, and intestinal tracts from day-of-hatch chicks were sampled for salmonellae contamination. Chicks from the same hatching trays were then put in isolators or floor pens and fed a pelleted corn-soybean ration for 1 week before external rinses and ceca from each chick were sampled for salmonellae. About 17% of eggshells, 21% of chick rinses, and 5% of intestines sampled at Day 0 were positive for salmonellae. No differences were observed between broiler hatcheries, but significant differences were seen between replications within hatching cabinets. Results from this study (Bailey et al., 1994) suggest a correlation between hatchery-acquired salmonellae and the production of potential seeder birds. No differences between eggshell and chick rinse samples were found (correlation = 0.81); therefore, eggshells were recommended as the best sample to determine hatchery salmonellae contamination.

6.7 Effect of Ventilation

Davies and Wray (1994) noted that in some hatcheries the main ventilation system drew air from areas where there was a potential source of *Salmonella* contamination, such as hatcher and chick area air exhaust ducts or the waste area where splashing of macerated egg and dead chick remains occurred. In areas where these hazards were present, *Salmonella* was consistently found in air intake ducts. In many cases, coarse filtration of air was used, but this was unlikely to restrict the access of small contaminated dust particles (Davies and Wray, 1994). The authors reported that, ideally, all hatcheries should be designed to draw in air from the opposite side of the building. Davies and Wray (1994) concluded that their investigations showed that in each of the key areas considered, it was possible to control *Salmonella* contamination.

Schoeni et al. (1995) compared the growth of *Salmonella* Enteritidis (SE), *Salmonella* Typhimurium, and *Salmonella heidelberg* inoculated into yolks and albumen at 4°C, 10°C, and 25°C. Regardless of whether 10^2 CFU/g or 10^4 CFU/g were inoculated into the yolk or albumen, populations of all strains increased by 3 $\log_{10}$ or more in number in 1 day when incubated at 25°C. Maximum numbers of *Salmonella* ranged from 10^8 to 10^{10} CFU/g (Schoeni et al., 1995). Another element of this study was to determine the potential for *Salmonella* in contaminated feces to establish itself in the interior of eggs by monitoring shell penetration. At 25°C, all three *Salmonella* strains penetrated the shell in 3 days, but at 4°C, only *Salmonella* Typhimurium was found in one membrane sample (Schoeni et al.,

1995). When hatchery conditions were simulated by incubating eggs at 35°C for 30 minutes followed by storage at 4°C, penetration of *Salmonella* into the egg-shell was enhanced. The authors also studied penetration when eggs were exposed to 10^4 to 10^6 CFU *Salmonella*/g feces. Increasing the inoculum to 10^6 CFU/g feces resulted in 50–75% of the contents of eggs to be contaminated by Day 1. The authors (Schoeni et al., 1995) concluded that the results indicated that SE, *Salmonella* Typhimurium, or *S. heidelberg* present in feces can penetrate to the interior of eggs and grow during storage.

6.8 Effect of Eggshell Parameters

In 2005, Messens et al. conducted a study in which egg weight, shell thickness, number of pores, cuticle deposition, and the ability of SE to penetrate the shell were determined for eggs from one layer flock through the entire production period. Penetration was assessed in this study by filling the eggs with a selective medium that allowed visualizing *Salmonella* growth on the inside of the shell and membrane complex. After each eggshell was inoculated with an average of 2.59 $\log_{10}$ CFU, the eggs were stored for up to 20 days at 20°C and 60% relative humidity. Messens et al. (2005) found that, on average, 38.7% of the eggshells were penetrated by *Salmonella*. Penetration of the eggs occurred most frequently on Day 3. The age of the hen did not significantly influence eggshell penetration, even though it did affect the physical parameters of the shell. The authors observed no correlations between any of the shell characteristics studied and the ability of *Salmonella* to penetrate the shell. The authors concluded that growth of SE on the shell is of major importance because shell contamination at 20 days of storage and SE penetration was highly correlated (Messens et al., 2005).

6.9 Effect of Long-Term Storage on Penetration

In a later study, Messens et al. (2006) determined the survival and penetration of SE inoculated on the eggshell and then stored for up to 20 days under real-world conditions (15–25°C and 45–75% relative humidity). The authors assessed penetration by emptying the egg contents and filling the eggs with a selective medium that allowed visualizing *Salmonella* growth on the inside of the shell and membrane complex. The survival of *Salmonella* on the eggshells was determined using viable counts and showed that numbers of surviving organisms decreased over time (Messens et al., 2006). Survival was inversely related to storage temperature and relative humidity. Although the average counts decreased over time, a limited proportion of shells carried high numbers of *Salmonella* at all storage conditions. After 20 days of storage, a similar percentage (44.7%) of eggshells became penetrated, irrespective of the storage conditions tested in this study. Messens et al. (2006)

reported that the higher the *Salmonella* shell contamination at the end of storage, the higher the probability that the eggshell was penetrated. *Salmonella* shell counts exceeding 4 $\log_{10}$ CFU yielded more than a 90% probability of eggshell penetration occurring (Messens et al., 2006).

Christensen et al. (1997) tracked the prevalence of *S. enterica* serovar Tennessee in broiler flocks in Denmark in the early months of 1994. Epidemiological studies showed that a single hatchery was involved in spreading the organism. In general, the authors reported that different strains of *S. enterica* ser. Tennessee had minor genotypic variation. Three different ribotypes were demonstrated when EcoRI was used for digestion of DNA. Two different types were obtained by the use of the restriction enzyme HindIII. Nine different plasmids and seven different plasmid profiles were demonstrated (Christensen et al., 1997). A 180-kb plasmid was, however, only demonstrated in isolates from broilers and the hatchery. Sixty-nine percent of the broiler isolates obtained during the period 1992 to 1995 harbored this plasmid, and 88% of the hatchery isolates contained a plasmid of the same size. An increased number of the broiler isolates (79%) contained this plasmid at the turn of 1994. Restriction enzyme analysis of the plasmid ensured that the plasmids from broilers and the hatchery were identical. By analyzing the cleaning and disinfection procedures and by sampling different control points in the hatchery, the authors were able to demonstrate that *S. enterica* ser. Tennessee had colonized areas of the hatchers that were protected from routine cleaning and disinfection. This supports the necessity for proper cleaning and sanitation of hatcheries specifically designed to remove biofilms that may remain on surfaces for long periods of time. The areas identified in these hatcheries that were colonized with *Salmonella* were cleaned and sanitized, which resulted in the elimination of *S. enterica* ser. Tennessee from the hatchers and a decrease in prevalence of *S. enterica* ser. Tennessee in broiler flocks during the following months (Christensen et al., 1997). This case represents a specific example of an epidemiological investigation leading to a cause, a solution being constructed, and a positive result being achieved.

In 1999, Byrd et al. determined the distribution of *Salmonella* serotypes from 5 commercial broiler hatcheries and 13 broiler farms. A total of 11 different *Salmonella* serotypes were isolated from hatcheries, with *Salmonella heidelberg* (9/30) and *Salmonella kentucky* (6/30) accounting for 50% of the total isolations. Of 700 chick paper pad tray liners sampled, regardless of lot (breeder flock source) or hatchery, 12% were positive for *Salmonella*. When 10 individual tray liners were cultured from individual lots (same breeder flock source), *Salmonella* was detected in 24/57 lots (42%). Multiple serotypes were simultaneously isolated from the same lot on three occasions (6%). Of the 21 lots that were serially sampled, the *Salmonella* serotype detected was different within lots eight times (38%) on at least one occasion of two or more sampling times. Of the 196 individual broiler houses sampled, 44 were positive for *Salmonella* (42%). Twelve different serotypes were isolated from broiler houses during this study. The serotypes isolated most frequently were *Salmonella heidelberg* (34/94) and *Salmonella kentucky* (22/94). These two serotypes accounted for 59.6% (56/94) of the total broiler house isolations. Of the 38 houses that were serially sampled, two or more serotypes were detected in the same broiler house on 20 occasions (53%). Of the 38 serially sampled houses (four or more times), a consistent *Salmonella* serotype was detected in 5 houses (13%). In only 5 of the 38 (13%) serially sampled houses did we fail to detect *Salmonella* on four or more samplings. No significant difference in *Salmonella* isolation frequency was observed between poultry houses using new or used litter. These data support previous findings indicating that paratyphoid *Salmonella* serotypes are prevalent in some broiler hatcheries and houses. Further, the observation of multiple serotypes simultaneously and serially isolated from the same breeder hatchery lots suggests that breeder flocks may be infected with more than one serotype, possibly providing a source for multiple serotype infections in progeny grower flocks.

Davies and Breslin (2004) reported that there are numerous routes and mechanisms by which the *Salmonella* may enter poultry flocks. These authors believed that concentration of the industry into a smaller number of larger enterprises has provided the opportunity for introduction of improved *Salmonella* control policies, but when failures in these systems occur, there are also higher risks of increased spread of the bacterium. Breeding flocks, feed mills, and hatcheries act as potential focal points of *Salmonella* contamination (Davies and Breslin, 2004). The authors reported that hatcheries may be responsible for wider distribution of infection originating from a single infected breeding flock. Also, hatcheries may become colonized by *Salmonella* that persists, often for many years (Christensen et al., 1997; Wilkins et al., 2002).

Davies and Breslin (2004) conducted a study in a hatchery in which 300 surface swab samples were taken with large gauze pads. The authors found significant *Salmonella* contamination of egg-handling areas indicative of ongoing infection in breeding flocks. Three of the hatcheries showed evidence of potential cross contamination during egg transfer. Hatcher incubator areas and chick-handling areas were contaminated in all of the hatcheries, and there was evidence of ineffective tray washing in four of the hatcheries (Davies and Breslin, 2004). *Salmonella* was detected in waste egg areas and in all of the hatchery buildings, and in some cases, contamination was also found in the main ventilation ducting and filters as well as in storerooms and vehicles. In this study, a wide variety of *Salmonella* serotypes and phage types were detected. The authors concluded that these studies have demonstrated both the dissemination of *Salmonella* contamination in the hatchery when infected breeding flocks have been present and persistence of the organism long after there is no current source of infection in hatchery eggs. Moreover, once a particular *Salmonella* serotype has become established in a hatchery, it is difficult to completely eliminate and becomes a source of contamination of chicks processed later in the building (Davies and Breslin, 2004).

6.10 Relationship of Contamination of the Hatchery and Flocks during Grow Out

In 2005, Pradhan et al. reported that contamination and penetration of salmonellae into hatching eggs may comprise an important link in the transmission of these bacteria to the grow-out operation, to processed carcasses, and eventually to the consumer. The authors found that incubated broiler eggs have an increase in internal bacterial loads between incubation and hatch.

6.11 Effect of Retained Yolk Sacs

Cox et al. (2006) conducted a study to determine the effect of retained yolk sacs on bacterial levels in full-grown broilers. The authors reported that, in the developing avian embryo, the main energy source is the yolk. Toward the end of the incubation period, the remaining yolk sac is internalized into the abdominal cavity. At hatch, the remaining yolk comprises 20% of the chick's body weight and provides the nutrients needed for maintenance (Cox et al., 2006). After hatching, chicks rapidly go from yolk dependence to the utilization of feed. Each carcass was aseptically opened and inspected for the presence of an unabsorbed yolk sac. Three to five carcasses containing a free-floating yolk sac (within the abdominal cavity) and the yolk stalk (without a yolk sac) and three to five carcasses containing an attached yolk and yolk stalk from each repetition were randomly selected and analyzed for *Salmonella* serovars. *Salmonella* serovars were found in 26% of the yolk stalks, 48% of the attached yolk sacs, and 23% of the free-floating yolk sacs, and the majority of *Salmonella* isolates were *Salmonella* Typhimurium; however, the significance of these bacterial reservoirs and carcass contamination during processing is yet to be determined (Cox et al., 2006).

6.12 Tracking Serotypes from the Hatchery to the the Carcasses

In Belgium in 2007, Heyndrickx et al. evaluated broiler flocks from the hatchery to the slaughterhouse using a multiple *Salmonella* typing approach (sero-, geno-, and phage types). The authors found that for 12 of the 18 flocks, there was no correlation between the serotypes found preharvest and those isolated from the feces in the transport coops and on the carcasses in the slaughterhouse; however, serotypes found in the coops were usually also found on the processed carcasses (Heyndrickx et al., 2007). The authors concluded that, most of the time, *Salmonella* strains that contaminate Belgian broiler carcasses do not predominate in the preharvest environment. This is interesting because it goes against the idea that vertical transmission is the primary problem with *Salmonella* colonization and contamination of poultry.

There is no question that the hatchery presents a serious opportunity for *Salmonella* on the external or internal surface of fertilized eggs to spread during incubation and hatching, thus increasing the prevalence in poultry populations. It is important for companies to employ intervention strategies to control this spread during hatching. These interventions are discussed in the next chapter.

References

Bailey J S, Blankenship L C, and Cox N A (1991), Effect of fructooliogsaccharide on *Salmonella* colonization of the chicken intestine, *Poultry Science*, 70, 2433–2438.

Bailey J S, Cox N A, and Berrang M E (1994), Hatchery acquired *Salmonella* in broiler chicks, *Poultry Science*, 73, 1153–1157.

Bailey J S, Cox N A, Blankenship L C, and Stern N J (1992), Hatchery contamination reduces the effectiveness of competitive exclusion treatments to control *Salmonella* colonization of broiler chicks, *Poultry Science*, 71(Suppl. 1), 6.

Bailey J S, Cox N A, Craven S E, and Cosby D E (2002), Serotype tracking of *Salmonella* through integrated broiler chicken operations, *Journal of Food Protection*, 65, 742–745.

Baker R C (1974), Microbiology of eggs, *Journal of Milk Food Technology*, 37, 265–268.

Bierer B W, Barnett B D, and Valentine H D (1961), Experimentally killing *Salmonella* Typhimurium on egg shells by washing, *Poultry Science*, 40, 1009–1014.

Blankenship L C, Bailey J S, Cox N A, Stern N A, Brewer R, and Williams O (1993), Two-step mucosal competitive exclusion flora treatment to diminish salmonellae in commercial broiler chickens, *Poultry Science*, 72, 1667–1672.

Brown M (1989), The scientists tell me ¼ Glutaraldehyde tested as an alternative sanitizer for egg disinfectant, *Dairy Food and Environmental Sanitation*, 9, 256.

Byrd J A, Deloach J R, Corrier D E, Nisbet D J, and Stanker L H (1999), Evaluation of *Salmonella* serotype distributions from commercial broiler hatcheries and grower houses, *Avian Diseases*, 43, 39–47.

Cason J A, Bailey J S, and Cox N A (1993), Location of *Salmonella typhimurium* during incubation and hatching of inoculated eggs, *Poultry Science*, 72, 2064–2068.

Cason J A, Cox N A, and Bailey J S (1994), Transmission of *Salmonella* Typhimurium during hatching of broiler chicks, *Avian Diseases*, 38, 583–588.

Christensen J P, Brown D J, Madsen M, Olsen J E, and Bisgaard M (1997), Hatchery-borne *Salmonella*-Enterica serovar Tennessee infections in broilers, *Avian Pathology*, 26, 155–168.

Corry J E L, Allen V M, Hudson W R, Breslin M F, and Davies R H (2002), Sources of *Salmonella* on broiler carcasses during transportation and processing: Modes of contamination and methods of control, *Journal of Applied Microbiology*, 92, 424–432.

Cox N A, Bailey J S, Mauldin J M, and Blankenship L C (1990), Research note: Presence and impact of *Salmonella* contamination in commercial broiler hatcheries, *Poultry Science*, 69, 1606–1609.

Cox N A, Bailey J S, Mauldin J M, and Blankenship L C, and Wilson J L (1991), Research note: Extent of salmonellae contamination in breeder hatcheries, *Poultry Science*, 70, 416–418.

Cox N A, Berrang M E, and Cason J A (2000), *Salmonella* penetration of egg shells and proliferation in broiler hatching eggs—A review, *Poultry Science*, 79, 1571–1574.

Cox N A, Richardson L J, Buhr R J, Northcutt J K, Fedorka-Cray P J, Bailey J S, Fairchild B D, and Mauldin J M (2006), Natural occurrence of *Campylobacter* species, *Salmonella* serovars, and other bacteria in unabsorbed yolks of market-age commercial broilers, *Journal of Applied Poultry Research*, 5, 551–557.

Davies R, and Breslin M (2004), Observations on the distribution and control of *Salmonella* contamination in poultry hatcheries, *British Poultry Science*, 45(Suppl. 1), S12-S14.

Davies R H, and Wray C (1994), An approach to reduction of *Salmonella* infection in broiler chicken flocks through intensive sampling and identification of cross-contamination hazards in commercial hatcheries, *International Journal of Food Microbiology*, 24, 147–160.

Frank J F, and Wright G W (1956), The disinfection of eggs contaminated with *Salmonella* Typhimurium, *Canadian Journal of Comparative Medicine*, 20, 406–410.

Gordon R F, Harry E G, and Tucker J F (1956), The use of germicidal dips in the control of bacterial contamination of the shells of hatching eggs, *Veterinary Record*, 68, 33–38.

Goren E, Dejong W A, Doornebal P, Bolder N M, Mulder R W A W, and Jansen A (1988), Reduction of salmonellae infection of broilers by spray application of intestinal microflora: A longitudinal study, *Veterinary Quarterly*, 10, 249–255.

Graves R C, and Maclaury D W (1962), The effects of temperature, vapor pressure, and absolute humidity on bacterial contamination of shell eggs, *Poultry Science*, 41, 1219–1225.

Gregory D W (1948), *Salmonella* infections of turkey eggs, *Poultry Science*, 27, 359–366.

Heyndrickx M, Herman L, Vlaes L, Butzler J P, Wildemauwe C, Godard C, and De Zutter L (2007), Multiple typing for the epidemiological study of the contamination of broilers with *Salmonella* from the hatchery to the slaughterhouse, *Journal of Food Protection*, 70, 323–334.

Humphrey T J, Baskerville A, Mawer S, Rowe B, and Hopper S (1989), *Salmonella* enteritidis phage type 4 from the contents of intact eggs: A study involving naturally infected eggs, *Epidemiology and Infection*, 103, 415–423.

Humphrey T J, Whitehead A, Gawler A H L, Henley A, and Rowe B (1991), Numbers of *Salmonella* Enteritidis in the contents of naturally contaminated hens eggs, *Epidemiology and Infection*, 106, 489–496.

Jones F T, Axtell R C, Rives D V, Scheideler S E, Tarver F R, Walker R L, and Wineland M J (1991a), A survey of *Salmonella* contamination in modern broiler production, *Journal of Food Protection*, 54, 502–507.

Jones F T, Axtell R C, Tarver F R, Rives D V, Scheideler S E, and Wineland M J (1991b), Environmental factors contributing to *Salmonella* colonization of chickens, in *Colonization Control of Human Bacterial Enteropathogens in Poultry*, ed L C Blankenship, Academic Press, San Diego, CA, 3–20.

Lahellec C, and Colin P (1985), Relationship between serotypes of salmonellae from hatcheries and rearing farms and those from processed poultry carcasses, *British Poultry Science*, 26, 179–186.

Lancaster J E, and Crabb W E (1953), Studies on disinfection of eggs and incubators. I. The survival of *Salmonella pullorum* and *typhimurium* on the surface of the hens egg and on incubator debris, *British Veterinary Journal*, 109, 139–148.

Maclaury D W, and Moran A B (1959), Bacterial contamination of hatching eggs, *Kentucky Agricultural Experiment Station Bulletin*, 665, 1.

Mellor D B, and Banwart G J (1965), Recovery of *Salmonella derby* from inoculated egg shell surfaces following sanitizing, *Poultry Science*, 44, 1244–1248.

Messens W, Grijspeerdt K, and Herman L (2005), Eggshell characteristics and penetration by *Salmonella enterica* serovar Enteritidis through the production period of a layer flock, *British Poultry Science*, 46, 694–700.

Messens W, Grijspeerdt K, and Herman L (2006), Eggshell penetration of hens eggs by *Salmonella* enterica serovar Enteritidis upon various storage conditions, *British Poultry Science*, 47, 554–560.

Muira S, Sato G, and Miyamae T (1964), Occurrence and survival of *Salmonella* organisms in hatcher chick fluff in commercial hatcheries, *Avian Diseases*, 8, 546–554.

Pradhan A K, Li Y, Swem B L, and Mauromoustakos A (2005), Predictive model for the survival, death, and growth of *Salmonella* Typhimurium in a broiler hatchery, *Poultry Science*, 84, 1959–1966.

Pritsker I Y (1941), Researches on the hatching qualities of eggs. II. Disinfection of egg shells under increased pressure within the egg, *Poultry Science*, 20, 102–103.

Rizk S S, Ayres J C, and Kraft A A (1966a), Disinfection of eggs artificially inoculated with salmonellae, *Poultry Science*, 45, 764–769.

Rizk S S, Ayres J C, and Kraft A A (1966b), Effect of holding condition on the development of salmonellae in artificially inoculated hen's eggs, *Poultry Science*, 45, 825–829.

Sauter E A, and Petersen C F (1974), The effect of egg shell quality on penetration by various salmonellae, *Poultry Science*, 53, 2159–2162.

Sauter E A, Petersen C F, Parkinson J F, and Steele E E (1977), Effects of pH on egg shell penetration by salmonellae, *Poultry Science*, 56, 1754–1755.

Schoeni J L, Glass K A, McDermott J L, Wong A C L (1995), Growth and penetration of *Salmonella* Enteritidis, *Salmonella* Heidelberg and *Salmonella* Typhimurium in eggs, *International Journal of Food Microbiology*, 24, 385–396.

Smeltzer T I, Orange K, Peel B, and Runge G (1979), Bacterial penetration in floor and nest box eggs from meat and layer birds, *Australian Veterinary Journal*, 55, 592–593.

Stokes J L, Osborne W W, and Bayne H G (1956), Penetration and growth of *Salmonella* in shell eggs, *Food Research*, 21, 510–518.

Walden C C, Allen I V F, and Trussell P C (1956), The role of the egg shell and shell membranes in restraining the entry of microorganisms, *Poultry Science*, 35, 1190–1196.

Wilkins M J, Bidol S A, Boulton M L, Stobierski M G, Massey J P, and Robinson-Dunn B (2002), Human salmonellosis associated with young poultry from a contaminated hatchery in Michigan and the resulting public health interventions, 1999 and 2000, *Epidemiology and Infection*, 129, 19–27.

Williams J E, and Dillard L H (1973), The effect of external shell treatments on *Salmonella* penetration of chicken eggs, *Poultry Science*, 52, 1084–1089.

Williams J E, Dillard L H, and Hall G O (1968), The penetration patterns of *Salmonella typhimurium* through the outer structures of chicken eggs, *Avian Diseases*, 12, 445–466.

Wright M L, Anderson G W, and Epps N A (1959), Hatchery sanitization, *Avian Diseases*, 3, 486.

Chapter 7

Hatchery Intervention

7.1 Introduction

Because the poultry hatchery has been shown to be a significant source of *Salmonella* spread, a number of studies have been conducted to attempt to prevent this spread during hatching. The first approach discussed is hatching egg disinfection.

7.2 Industry Survey

In 1997, Cox et al. conducted a study to compare the percentage of salmonellae-positive samples from 1990 to 1995 to determine how well the poultry industry was doing in terms of lowering the presence of *Salmonella*. *Salmonella* positives dropped from 75.4% in 1990 to 26.0% in 1995. The authors reported that these impressive reductions were most probably influenced by one or a combination of the following: (1) reduction of salmonellae contamination in breeder flocks, (2) improvement in the hatchery sanitation program (more attention to detail), (3) use of more effective chemicals for on-farm procedures and in hatchery sanitization, (4) more diligent cleaning and changing of nest material on the breeder farm, (5) improved ventilation in the hatchery, and (6) an assortment of other factors, such as rodent control and biosecurity at the breeder farm (Cox et al., 1997).

In another study in 2001, Cox et al. collected eggshell fragments, paper pads from chick boxes, and fluff samples from three commercial primary breeder hatcheries and analyzed for the presence and level of salmonellae with identical laboratory methods in 1991 and 1998. Overall, 29 of 180 samples (16.1%) from the three hatcheries in 1998 were contaminated with salmonellae, whereas in 1991, 11.1% of the overall samples were salmonellae positive (Cox et al., 2001). Salmonellae were detected in 1.7% of eggshell fragments, 1.7% of fluff samples, and 48% of the paper pad samples in 1998, whereas 15.2%, 4.5%, and 12% of these type samples, respectively, were salmonellae positive in 1991. The authors reported that, although the percentage of positive samples was slightly higher in 1998 than 1991, from an enumeration standpoint, the salmonellae contamination in primary breeder hatcheries seemed to have improved in the 7 years. In 1998, less than 4% of the positive samples had high levels of salmonellae, whereas 36% of the positive samples in 1991 had high numbers of salmonellae. Thus, it seemed that the industry was able to reduce the number of samples with high concentrations of *Salmonella* over the 7-year period. It is unclear how this reduction would have an impact on colonization spread during grow out.

7.3 Efficacy of Sanitizers

7.3.1 Hydrogen Peroxide

Sheldon and Brake (1991) researched the effectiveness of hydrogen peroxide (H_2O_2) at different concentrations to disinfect broiler hatching eggshell surfaces while maintaining percentage hatch. A concentration of 5% H_2O_2 was required on eggshells to disinfect the shell surfaces (approximately 5 $\log_{10}$ reduction). Most H_2O_2 is sold as a 3% solution, so concentrated H_2O_2 would be needed for this application. The authors reported that hatchability of fertile eggs from a 44-week-old flock was significantly increased by 2% after spraying the eggs with 5% H_2O_2 in comparison to untreated controls (Sheldon and Brake, 1991). Moreover, the number of contaminated eggs and "early-dead" embryos were significantly reduced in the H_2O_2-treated eggs. The authors concluded that the results demonstrated that H_2O_2 compared favorably to formaldehyde as a hatching egg disinfectant without adversely affecting hatching potential.

7.3.2 Ultraviolet Light, Ozone, or H₂O₂

Bailey et al. (1996) conducted trials to evaluate the efficacy of hatcher air sanitation utilizing ultraviolet (UV) light, ozone, or H_2O_2 on bacterial populations, the spread of *Salmonella*, and hatchability of broiler eggs. The UV light (254 nm, 146 mW/second) and ozone (0.2 or 0.4 ppm) treatments were continuously applied through the last 3 days of hatch, whereas the H_2O_2 treatment (2.5%) was administered for 1 or 2 minutes of each 10 minutes at rates of 500 or 100 mL/hour. Hatchability was reduced by sanitizing treatments when compared with the untreated control (94 vs. 95.6%), but this difference was not significant (Bailey et al., 1996). When compared to controls, all sanitizing treatments reduced 75–99% of the total bacteria, Enterobacteriaceae, and *Salmonella* in the hatching cabinet air samples. H_2O_2 reduced bacteria more than ozone or UV light. Only hydrogen peroxide significantly reduced *Salmonella* levels on eggshell fragments (Bailey et al., 1996). The authors reported that these reductions on the eggshells led to significant reductions in the number of *Salmonella*-positive chicks when ozone and H_2O_2 were used. H_2O_2 was able to significantly reduce *Salmonella* colonization in chicken ceca. Bailey et al. (1996) concluded that the spread of bacteria can be effectively reduced in the hatching cabinet by air sanitization using UV light, ozone, and hydrogen peroxide without depressing hatchability.

7.3.3 H_2O_2 or Timsen

Cox et al. (2002) immersed fertile hatching eggs into varying levels (10^5 to 10^7/egg) of *Salmonella* Typhimurium. The groups tested included untreated (control), water-treated, hydrogen peroxide- (1.5%) treated, or Timsen- (n-alkyl dimethyl benzyl ammonium chloride as a commercial bactericide-fungicide) treated hatching eggs. Hydrogen peroxide was superior to Timsen as an egg treatment to eliminate artificially inoculated *Salmonella* from fertile eggs, but one-third of the treated eggs remained *Salmonella* positive (Cox et al., 2002). The authors concluded that it is difficult to eliminate *Salmonella* that contaminate fertile hatching eggs, and until a more effective system or process is devised and commercially implemented, *Salmonella* will continue to pass from one generation to the next through the fertile egg.

7.4 Electrostatic Application of Sanitizers during Incubation and Hatching

While a number of companies currently use fogging systems or formaldehyde gas to disinfect the hatchery environment, these systems are problematic in that they are expensive (due to the amount of sanitizer that must be used), are hazardous to the health of employees (glutaraldehyde and formaldehyde are noxious), may have a negative impact on chick quality and hatchability, and may be corrosive to equipment. Part of the problem involved in disinfecting hatching eggs is that they are tightly packed into incubators or into hatching trays (Figure 7.1).

This limits the ability of poultry hatchery operators to apply sanitizers to the shells of the eggs. A new technology has been developed that has been shown to be effective for applying sanitizers to fertile hatching eggs.

7.4.1 Development of Electrostatic Spraying

Electrostatic spraying was developed over two decades ago, and its most common use has been for applying pesticides to row crops. In 1978, Dr. Law of the University

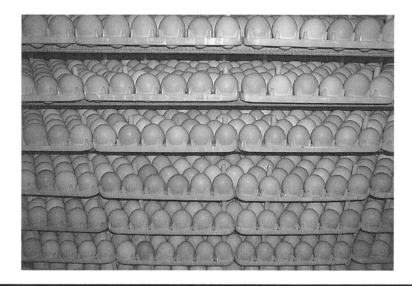

Figure 7.1   Hatching eggs tightly packed into the setter.

of Georgia developed an electrostatic spray-charging system using air atomization that could be used to achieve a sevenfold increase in spray deposition over conventional application methods. In a study conducted in 1981, Dr. Law found a 1.6- to 24-fold increase in deposition. This is because surfaces such as eggs, equipment, and walls have a native positive charge. As high-pressure air and sanitizer are forced through a small aperture in the electrostatic spray nozzle (Figure 7.2), the air shears the sanitizer into tiny droplets (approximately 30 mm in diameter). These droplets are then exposed to an electrical charge as they exit the nozzle head. This transfers a negative charge to the sanitizer particle, which then has a particular affinity for the surfaces in the area, such as eggs or equipment.

7.4.2 Electrostatic Spraying Saves Money and Material

Because the deposition of sanitizer to the surface being treated is so much more efficient, much less sanitizer is required to result in the same bacterial disinfection rate when compared to commonly used commercial foggers or sprayers. In 1983, Herzog et al. demonstrated this effect and showed that insect control on cotton plants was equal to or better than conventional spray application using only one-half the amount of insecticide.

The circular, flat metal ring is where the charge is applied to the sanitizer as the air/sanitizer mixture is vortexed down the center of the nozzle head and out the end of the nozzle. Figures 7.3 and 7.4 clearly demonstrate the difference between using a traditional sprayer and an electrostatic sprayer. The deposition is many orders of magnitude more effective using electrostatic spraying.

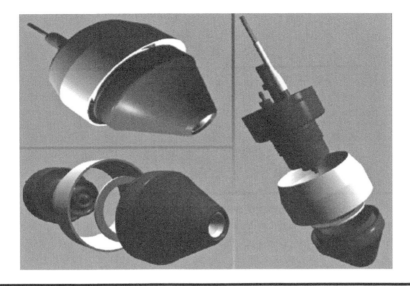

Figure 7.2   A diagram of the electrostatic spraying nozzle. (Figure used with permission from Bruce Whiting of Electrostatic Spraying Systems, Inc., Watkinsville, GA.)

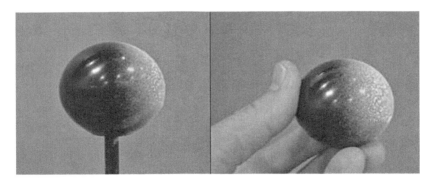

Figure 7.3   A black ball sprayed with a powder solution using a traditional mister spraying nozzle. (Figure used with permission from Jack Mathis, Maxspray.com.)

Figure 7.5 shows the nozzle setup in a hatchery plenum. Figure 7.6 shows sanitizer being applied within a hatching cabinet and in the plenum area of a hatchery.

Figures 7.3 and 7.4 demonstrate the efficacy of electrostatic spraying on coating a black ball (model for an egg) with a disinfectant. Table 7.1 indicates that disinfecting the eggs with electrolyzed oxidizing (EO) water in a study funded by the U.S. Poultry and Egg Association had no impact on hatchability. Table 7.2 shows the data obtained during that study for prevention of colonization of chickens using electrostatic application of EO water during the hatching process.

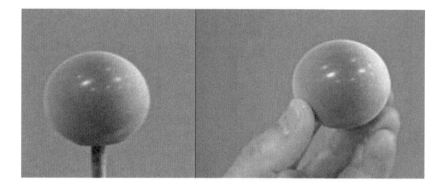

Figure 7.4   A black ball sprayed with a powder solution using an electrostatic spraying nozzle. (Figure used with permission from Jack Mathis, Maxspray.com.)

Figure 7.5   Nozzle setup in a hatchery plenum.

As is demonstrated in the photos, the sanitizer is sprayed as a very fine fog and in a short period of time completely disappears. This fog completely covers every surface within the hatching cabinet, including eggs. Complete coverage has been demonstrated using fluorescent dye sprayed onto surfaces. After spraying, the area can be evaluated using a black light, and even the most difficult-to-reach spaces are completely covered.

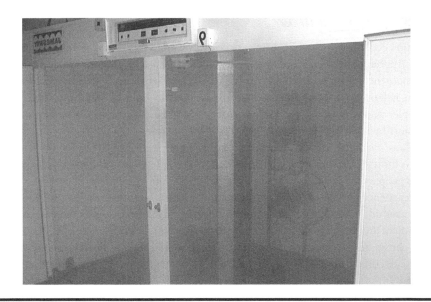

Figure 7.6   Electrostatic spraying of sanitizer in hatching cabinet.

Table 7.1   Results for Hatchability of Commercial Broiler-Breeder Chicks from Hatching Eggs Treated Electrostatically with Tap Water or EO Water during Hatch under Commercial Conditions

	Normal Hatch[a]	*Tap Water Treated*	*EO Water Treated*
Hatchability	85%	82%	79%
n		15,000	15,000

[a] Fertile eggs used in this study for the EO water treatment were older and expected hatchability was lower than the normally expected hatch.

Table 7.2   Results for Hatchability of Chicks from Hatching Eggs Treated Electrostatically with Tap Water or EO Water during Hatch under Research Conditions at the University of Georgia Poultry Research Center

	Normal Hatch	*Tap Water Treated*	*EO Water Treated*
Hatchability	92%	93%	93%
n		160	160

7.4.3 Type of Sanitizer

A major consideration when using electrostatic spraying is the type of sanitizer used. Applying an electrical charge or atomization has the potential to completely eliminate the killing power of some sanitizers. Thus, when using electrostatic spraying, it is best to evaluate the sanitizer to be used in light of this limitation. For example, oxidative sanitizers, such as chlorine, can be reduced in concentration by 90% just by spraying them using electrostatic sprayers.

Many currently used sanitizers are objectionable for various reasons. Formaldehyde is difficult to work with and presents a worker safety hazard. Formaldehyde gas burns people's lungs and eyes when they are exposed to it, and many have compared the experience to that of being exposed to tear gas. Glutaraldehyde is also difficult to be around. Concentrated hydrogen peroxide, while effective, may corrode some equipment and is irritating to the lungs when breathed. It would seem that exposing baby chicks on day of hatch to these chemicals would be disadvantageous if other chemicals could be used that would be as effective at eliminating pathogenic bacteria.

7.4.4 Quaternary Ammonium and H₂O₂ Applied Using Electrostatic Spraying

In 1993, Russell conducted research to compare the efficacy of BioSentry 904 (quaternary ammonium mixture) and Biox H (concentrated hydrogen peroxide) applied using an electrostatic spraying system for killing populations of bacteria that are of concern to the poultry industry. Populations of pathogenic bacteria such as *Salmonella* Enteritidis (SE), *Staphylococcus aureus* (SA), and *Listeria monocytogenes* (LM) and the indicator bacterium *Escherichia coli* (*E. coli*) were appled to eggs and allowed to attach for 1 hour. BioSentry 904 completely eliminated all SA on 100%, 93%, and 60% of eggs, and BioxH eliminated all SE on 100%, 93%, and 93% of eggs in Repetitions 1, 2, and 3, respectively. BioSentry 904 completely eliminated all SE on 100%, 87%, and 100% of eggs and BioxH eliminated all SA on 100%, 100%, and 80% of eggs in Repetitions 1, 2, and 3, respectively. BioSentry 904 completely eliminated all LM on 100% of eggs and BioxH eliminated all LM on 93%, 87%, and 73% of eggs in Repetitions 1, 2, and 3, respectively. BioSentry 904 completely eliminated all *E. coli* on 93%, 93%, and 60% of eggs and BioxH eliminated all *E. coli* on 93% of eggs in Repetitions 1, 2, and 3, respectively. Russell (1993) found that BioSentry 904 and BioxH were extremely effective when used in conjunction with electrostatic spraying for eliminating pathogenic and indictor populations of bacteria from eggshell surfaces.

7.4.5 Electrolyzed Oxidative Water Applied Using Electrostatic Spraying

In another study, Russell (2003) conducted a study to determine if electrostatic spraying of EO water was effective in eliminating *Salmonella* spp. from fertile hatching eggs and to evaluate if this reduction carried through to a significant reduction in colonization of chickens during the grow-out process. It was observed in these studies that electrostatic application of EO water completely removed the dust, fluff, and dander from the air, upper surfaces of the hatching cabinet, and the eggs. It is believed that by charging the EO water using electrostatic spraying the dust and dander were also charged and fell to the floor, away from the eggs and chicks (Russell, 2003).

Results for *Salmonella typhimurium* prevalence in the lower intestines of broiler chickens from hatching eggs treated electrostatically with tap water or EO water during hatch are presented in Table 7.3. These results were extremely encouraging in that 65–95% (Replicates 1 and 2, respectively) of the chickens were colonized when only tap water was used to treat the fertile hatching eggs, indicating that the method for inducing colonization was appropriate; however, for electrostatically treated eggs using EO water, *Salmonella* was only able to colonize 1 chicken of 40 tested over two repetitions under actual grow-out conditions. This research has tremendous industrial application because many of the companies that are experiencing failures due to high *Salmonella* prevalence at the poultry plant are receiving flocks of birds that are 80–100% positive for *Salmonella* as they enter the plant. It would seem logical to suppose that if the number of chickens in the field that are colonized with *Salmonella* could be reduced to the levels observed in this study, the industry would be able to meet the *Salmonella* performance standard required by the U.S. Department of Agriculture, Food Safety Inspection Service (USDA-FSIS). This research described a method that should have tremendous value to the poultry industry for reducing *Salmonella* in flocks arriving to the processing plant, which, according to our research, will translate directly into lower numbers of processed carcasses that are positive for *Salmonella*. Moreover, the electrostatic spraying system is not expensive to incorporate into a commercial hatchery. In addition, the EO water is economical to produce and is so nontoxic as to be potable. This water does not degrade equipment and does not present an environmental hazard when discharged.

Table 7.3 Results for *Salmonella typhimurium* Prevalence (%) in the Lower Intestines, Ceca, or Cloacae of Broiler Chickens from Hatching Eggs Treated Electrostatically with Tap Water or EO Water during Hatch

Treatment	Repetition 1	Repetition 2
Tap water control	65%	95%
EO water treated	0%	5%
n	20	20

7.4.6 Ionized Reactive Oxygen

In 2005, Higgins et al. evaluated the effect of ionized reactive oxygen species (ROS) created using binary ionization technology (BIT) for disinfection of broiler carcasses and table eggs and treatment of fertile eggs. Previous research has indicated that BIT creates a high concentration of ROS that lyse bacterial cells on contact. After inoculation of table eggs with 6.8×10^8 CFU of SE, the authors recovered SE from 95% fewer eggs following enrichment and found significantly fewer (7.77 and 7.41 $\log_{10}$ reduction) colony-forming units recovered from eggs treated with BIT compared with nontreated control eggs (Higgins et al., 2005). The authors also tested hatchability and determined that there were no significant effects of BIT on the hatchability (of the total set) of treated eggs as compared with nontreated control eggs; however, there was a slight numerical increase in hatchability, between 5% and 10% in two trials. The conclusion was that application of BIT to carcasses and table eggs could reduce contamination with pathogens, and that the application to fertile eggs may not have effects on hatchability of eggs (Higgins et al., 2005).

7.4.7 Suggestions for Elimination of Salmonella in the Hatchery

1. Install a disinfectant fogging system or electrostatic spraying system in the hatchery plenum, setters, and hatchers that is linked to a timer system.
2. Spray a disinfectant (hydrogen peroxide, EO water, quaternary ammonium, or chlorine dioxide) every 30 minutes during setting and hatching to prevent cross contamination.
3. Thoroughly clean and sanitize setters and hatchers regularly using documented sanitation standard operating procedures (SSOPs).
4. Regularly monitor eggshell fragments, chick paper pads, and chick dander from the bottom of the hatching cabinet for *Salmonella*.

7.5 In Ovo Competitive Exclusion in the Hatchery

Exposure of baby chicks to salmonellae in the hatchery and hatchery environment limits the effectiveness of a competitive exclusion (CE) culture treatment to colonize the intestine of the chick and thus prevent *Salmonella* colonization (Cox et al., 1992). These authors used a technique to introduce the CE culture before chicks are exposed to salmonellae by placing the CE culture in ovo to unhatched embryos. An undefined, anaerobically grown CE culture, derived from cecal contents of healthy adult chickens, was diluted 1:1,000 or 1:1,000,000 and inoculated either into the air cell or beneath the inner air cell membrane of 17-day-old incubating hatching eggs (Cox et al., 1992). In this study, treated chicks were more resistant than untreated chicks to different challenge levels of *Salmonella* Typhimurium, indicating that it may be possible to initiate protection of chicks to salmonellae challenge prior to hatching into a contaminated environment.

In another study, Bailey et al. (1998) evaluated the potential of *Salmonella* contamination in hatching cabinets to (1) generate seeder chicks that will spread *Salmonella* to other chickens during grow out and (2) interfere with the efficacy of CE treatments. Hatchery-generated seeder chicks were produced by hatching in the same hatcher with eggs inoculated with *Salmonella* Typhimurium (10^5 to 10^6). When hatchery-generated seeder chicks were stocked in floor pens at a 1:10 ratio with uncontaminated contact chicks, the pen environment became greater than 50% contaminated (Bailey et al., 1998). When eggs were placed into a hatching cabinet with only 1 to 3 inoculated eggs per 200-egg hatching cabinet, 98% of uninoculated chicks were intestinally colonized with *Salmonella* after the birds were held 1 week in isolation cabinets. This hatchery-acquired *Salmonella* substantially reduced the effectiveness of subsequent CE treatments to prevent *Salmonella* colonization of the young chicks. These studies demonstrated that control of *Salmonella* in hatching cabinets is critically important for control of *Salmonella* in broilers (Bailey et al., 1998).

Other scientists studied a commercial preparation of CE culture composed of normal avian gut flora (NAGF). A culture of these bacteria was aerosolized for an extended period over turkey hatching eggs during pipping and hatching to examine any protective effects against natural exposure to salmonellae (Primm et al., 1997). Turkey hatching eggs, produced by salmonellae-infected breeder flocks and hatched in a commercial hatchery with a history of salmonellae contamination, were investigated. In both of the trials conducted, turkey poults were exposed to *Salmonella montevideo* during hatching. After 7 days of growth, treated poults were culture negative for salmonellae, and control poults were culture positive for salmonellae. The authors concluded that these studies justify further critical evaluation of the protective effects of prolonged aerosolization of NAGF during pipping and hatching against salmonellae colonization in turkey poults (Primm et al., 1997). Unfortunately, the U.S. Food and Drug Association does not allow the poultry industry in the United States to use undefined gut flora preparations.

Edens et al. (1997) stated that use of in ovo CE agents is feasible for both chickens and turkeys; however, there are many pitfalls that await the use of in ovo application of CE agents, including the use of non-species–specific intestinal microbes and the use of harmful proteolytic, gas-producing and toxin-producing intestinal microbes. Of the potential CE agents that have posthatch application, only *Lactobacillus reuteri* has been shown to be safe and effective in terms of not affecting hatchability and in having a prolonged effect in the hatched chick or poult. *Lactobacillus reuteri* administration in ovo increases its rate of intestinal colonization and decreases the colonization of *Salmonella* and *E. coli* in both chicks and poults (Edens et al., 1997). The authors added that mortality due to in-hatcher exposure to *E. coli* or *Salmonella* was reduced with in ovo application of *L. reuteri*. Edens et al. (1997) also measured physical parameters within the intestines of the birds and found that *L. reuteri* both in ovo and ex ovo increased villus height and crypt depth. The authors concluded that these approaches may be useful for controlling *Salmonella* colonization.

Many researchers have been trying to find defined cultures of bacteria that may be used in CE applications. Tellez et al. (2001) conducted a study to determine the effect of an avian-specific probiotic combined with antibodies specific for *Salmonella* Enteritidis, *Salmonella* Typhimurium, and *Salmonella heidelberg* on the cecal colonization and organ invasion of *Salmonella* Enteritidis in the broiler chicken. The treatment group was defined as chicks spray-vaccinated with a product named Avian Pac Plus at the hatchery and for the first 3 days after placement. Another intermediate treatment was given at 10 and 14 days, 2 days prior to vaccination and 2 days postvaccination, respectively (Tellez et al., 2001). A final treatment was given to the birds 3 days before slaughter. The authors defined the control group as chicks not given Avian Pac Plus at any time. All chickens were orally inoculated with 0.25 mL of *Salmonella* Enteritidis that contained 4×10^7 CFU/mL. The authors evaluated cecal colonization and organ invasion for *Salmonella* Enteritidis on Days 0, 1, 3, 7, 10, 17, 24, 31, 38, and 41. Tellez et al. (2001) concluded that the probiotic-treated group had a significantly lower concentration of *Salmonella* Enteritidis cecal colonization at all days after Day 0, when compared to the nontreated, control group.

Other scientists reported that undefined Nurmi-type cultures (NTCs) have been used successfully to prevent *Salmonella* colonization in poultry for decades (Waters et al., 2006). Such cultures are derived from the cecal contents of specific-pathogen–free birds and administered via drinking water or spray application onto eggs in the hatchery. These cultures consist of many nonculturable and obligately anaerobic bacteria. Due to their undefined nature, it is difficult to obtain approval from regulatory agencies to use these preparations as direct-fed microbials for poultry. Waters et al. (2006) set out to identify the bacteria in these undefined cultures. A number of *Lactobacillus* spp. were found in these cultures, including *L. fermentum, L. pontis, L. crispatus, L. salivarius, L. casei, L. suntoryeus, L. vaginalis, L. gasseri, L. aviaries, L. johnsonii, L. acidophilus,* and *L. mucosae,* in addition to a range of unculturable lactobacilli. While NTCs are successful due to their complexity, the presence of members of *Lactobacillus* spp. among other probiotic genera in these samples possibly lends to the success of the NTCs as probiotics or CE products in poultry over the decades (Waters et al., 2006). These findings are essential in understanding why some defined cultures of *Lactobacillus* do not work effectively, while complex mixtures of these bacteria have been shown to be effective.

7.6 Treating Hatcher Air to Reduce Fomites

Mitchell (1998) reported that hatching cabinets are known to be one of the primary sources for *Salmonella* contamination of poultry. A considerable amount of dust and dander fomites are generated during the hatching process from the time of pipping on Day 20 through final hatching on Day 21. The dust is formed when the eggshells break and when feather particles from the new chicks move around in the cabinet (Mitchell, 1998). Two configurations of negative air ionizers were evaluated for their ability to remove inhalable particulates from air in a poultry hatching cabinet to determine their potential for reducing airborne disease transmission. The treatments were applied to ambient dust that was found to have high particle counts in the inhalable range. The author made measurements with a laser particle counter. Significant reductions in particle counts were achieved with the six-bar Room Ionizer System (RIS), which removed particles with efficiencies that averaged 92.9% for particles up to 10 mm and 90.8% for particles 10 mm and larger (Mitchell, 1998). The author concluded that the efficacy of the ionizers for removing dust and knowledge that most airborne microorganisms are found on particles larger than 3 mm suggest that this type of system has significant potential for reducing airborne transmission of disease in hatching cabinets.

Another approach tested by Mitchell et al. (2002) involved electrostatic charging of particles in enclosed spaces. The authors reported that this approach is an effective means of reducing airborne dust (Figures 7.7 and 7.8).

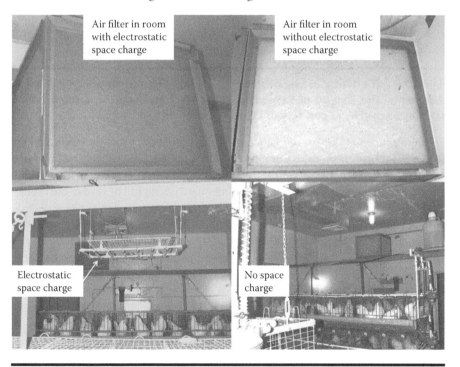

Figure 7.7   The effect of electrostatic space charging the air (filter on the left) versus no space charging on the amount of dust in the air in a room full of hatching eggs. (Figure courtesy of Dr. Bailey Mitchell of Southeast Poultry Research Laboratory USDA-Agricultural Research Service, Athens, GA.)

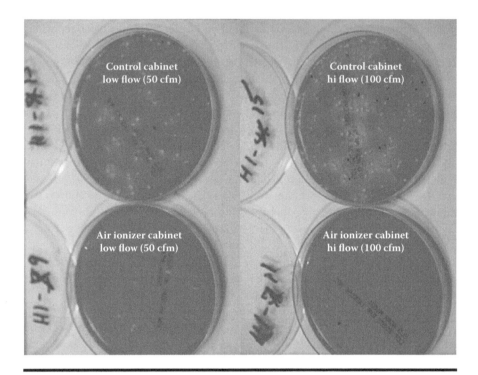

Figure 7.8 **The effect of electrostatic space charging the air versus no space charging on the amount of** *Salmonella* **collected from the air in a room full of hatching eggs. (Figure courtesy of Dr. Bailey Mitchell.)**

Dust generated during the hatching process has been strongly implicated in *Salmonella* transmission, which complicates the cleaning and disinfecting processes for hatchers. Mitchell et al. (2002) placed the electrostatic space charge system (ESCS) in a hatching cabinet that was approximately 50% full of 18-day-old broiler hatching eggs. The ESCS operated continuously to generate a strong negative electrostatic charge throughout the cabinet through hatching, and dust was collected in grounded trays containing water and a degreaser (Mitchell et al., 2002). An adjacent hatching cabinet served as an untreated control. Air samples from hatchers were collected daily, and sample chicks from each hatcher were grown out to 7 days of age for cecal analysis in three of the trials. The ESCS significantly reduced aerobic plate counts (APCs) and Enterobacteriaceae (ENT) by 85–93%. Dust concentration was also significantly reduced during the preliminary trials, with an average reduction of 93.6%. Most important, the number of *Salmonella* per gram of cecal contents in birds grown to 7 days of age was significantly reduced by an average of 3.4 $\log_{10}$ CFU/g. The authors concluded that ionization technology is relatively inexpensive and could be used to reduce airborne bacteria and dust within the hatching cabinet (Mitchell et al., 2002).

References

Bailey J S, Buhr R J, Cox N A, and Berrang M E (1996), Effect of hatching cabinet sanitation treatments on *Salmonella* cross-contamination and hatchability of broiler eggs, *Poultry Science*, 75, 191–196.

Bailey J S, Cason J A, and Cox N A (1998), Effect of *Salmonella* in young chicks on competitive exclusion treatment, *Poultry Science*, 77, 394–399.

Cox N A, Bailey J S, Berrang M E, and Mauldin J M (1997), Diminishing incidence and level of salmonellae in commercial broiler hatcheries, *Journal of Applied Poultry Research*, 6, 90–93.

Cox N A, Bailey J S, Blankenship L C, and Gildersleeve R P (1992), In ovo administration of a competitive exclusion culture treatment to broiler embryos, *Poultry Science*, 71, 1781–1784.

Cox N A, Berrang M E, and Mauldin J M (2001), Extent of salmonellae contamination in primary breeder hatcheries in 1998 as compared to 1991, *Journal of Applied Poultry Research*, 10, 202–205.

Cox N A, Mauldin J M, Kumararaj R, and Musgrove M T (2002), Ability of hydrogen peroxide and Timsen to eliminate artificially inoculated *Salmonella* from fertile broiler eggs, *Journal of Applied Poultry Research*, 11, 266–269.

Edens F W, Parkhurst C R, Casas I A, and Dobrogosz W J (1997), Principles of ex-ovo competitive exclusion and in ovo administration of *Lactobacillus reuteri*, *Poultry Science*, 76, 179–196.

Herzog G A, Law S E, Lambert W R, Seigler W E, and Giles D K (1983), Evaluation of an electrostatic spray application system for control of insect pests in cotton, *Journal of Ecological Entomology*, 76, 637–640.

Higgins S E, Wolfenden A D, Bielke L R, Pixley C M, Torres-Rodriguez A, Vicente J L, Bosseau D, Neighbor N, Hargis B M, and Tellez G (2005), Application of ionized reactive oxygen species for disinfection of carcasses, table eggs, and fertile eggs, *Journal of Applied Poultry Research*, 14, 716–720.

Law S E (1978), Embedded-electrode electrostatic induction spray charging nozzle: Theoretical and engineering design, *Transactions of the American Society of Agricultural Engineers*, 12, 1096–1104.

Mitchell B W (1998), Effect of negative air ionization on ambient particulates in a hatching cabinet, *Applied Engineering in Agriculture*, 14, 551–555.

Mitchell B W, Buhr R J, Berrang M E, Bailey J S, and Cox N A (2002), Reducing airborne pathogens, dust and *Salmonella* transmission in experimental hatching cabinets using an electrostatic space charge system, *Poultry Science*, 81, 49–55.

Primm N D, Vance K, Wykle L, and Hofacre C L (1997), Application of normal avian gut flora by prolonged aerosolization onto turkey hatching eggs naturally exposed to *Salmonella*, *Avian Diseases*, 41, 455–460.

Russell S M (1993), Effect of sanitizers applied by electrostatic spraying on pathogenic and indicator bacteria attached to the surface of eggs, *Journal of Applied Poultry Research*, 12, 183–189.

Russell S M (2003), Prevention of *Salmonella* colonization of chickens: Electrolyzed oxidative acidic water. University of Georgia, publication 81313. http://www.caes.uga.edu/publication/PubDetail.cfm?pk_id=7467.

Sheldon B W, and Brake J (1991), Hydrogen peroxide as an alternative hatching egg disinfectant, *Poultry Science*, 70, 1092–1098.

Tellez G, Petrone V M, Escorcia M, Morishita T Y, Cobb C W, Villasenor L, and Promsopone B (2001), Evaluation of avian-specific probiotic and *Salmonella* enteritidis-, *Salmonella* Typhimurium-, and *Salmonella heidelberg*-specific antibodies on cecal colonization and organ invasion of *Salmonella* Enteritidis in broilers, *Journal of Food Protection*, 64, 287–291.

Waters S M, Murphy R A, and Power R F G, 2006. Characterisation of prototype Nurmi cultures using culture-based microbiological techniques and PCR-DGGE, *International Journal of Food Microbiology*, 110, 268–277.

Chapter 8

Salmonella Transfer during Grow Out

8.1 Introduction

The process of growing broiler chickens is replete with opportunities for *Salmonella* to be transferred from the environment to the chickens or from chicken to chicken. Cox et al. (1996) suggested that the newly hatched chick may be exposed to significant levels of salmonellae from an assortment of sources, such as the hatching cabinet, hatchery environment, and broiler house. Once salmonellae reach the ceca of a young chick, they may attach to the epithelium and multiply to high numbers in a relatively short period. In this situation, the young chick may excrete large numbers of salmonellae in its cecal droppings, a situation that will result in the contamination of other birds in the broiler house (Cox et al., 1996). A study was conducted in which salmonellae were introduced into the day-of-hatch chick through an assortment of body openings (mouth, cloaca, eye, nasal passage, and navel) to determine which of the openings would potentially result in the production of seeder birds. The production of seeder birds readily occurred when salmonellae were introduced via the mouth, cloaca, eye, and nasal passage (Cox et al., 1996). The data from this study suggest that potential salmonellae seeder birds can develop from contamination of various body openings in the newly hatched baby chick, emphasizing the need to control salmonellae in breeder flocks, hatcheries, and broiler houses.

Figure 8.1 Adult darkling beetle. (Figure used with permission from Tommy Powell. http://www.pestproducts.com/darkling-beetle.htm.)

8.2 Role of Insects

Insects such as darkling beetles (Figure 8.1) and flies can spread *Salmonella* around the grow-out house. Jones et al. (1991) found that insects were responsible for mechanically carrying *Salmonella* throughout the houses. Evidence also exists that *Salmonella* spp. may survive in the intestinal tracts of insects. Research has shown that houseflies and cockroaches may feed on contaminated feces and become reservoirs of *Salmonella* spp. (Jones et al., 1991). Klowden and Greenberg (1976, 1977) studied cockroaches challenged with 10^6 *Salmonella* Typhimurium and reported a colonization rate of only 3% of their experimental population and no replication within the gut, presumably due to the resident gut microflora of the insect. However, Jones et al. (1991) and Kopanic et al. (1994) suggested that insects and cockroaches, respectively, may be vectors for the transmission of *Salmonella*.

8.3 Relationship of Contamination of Birds and *Salmonella* on Carcasses

Rasschaert et al. (2008) studied successively slaughtered poultry flocks to determine the relationship between gastrointestinal colonization of the birds with *Salmonella* and contamination of the carcasses after slaughter. Although only 7 (13%) of the broiler flocks were colonized with *Salmonella* at slaughter, the carcasses of 31

(55%) broiler flocks were contaminated after slaughter, indicating an increase of *Salmonella* prevalence throughout the processing operation. With regard to the layer and breeder flocks, 11 (69%) flocks were colonized in their gastrointestinal tracts, but after slaughter, carcasses of all flocks were contaminated (Rasschaert et al., 2008). The authors reported that *Salmonella* prevalence at the farm did not always correlate to prevalence at slaughter. In addition, *Salmonella* isolated from positive flocks did not result in the same *Salmonella* pulsed-field gel electrophoresis (PFGE) types isolated from the gastrointestinal tract. Interestingly, when *Salmonella*-negative flocks were slaughtered, some became positive for *Salmonella*. Rasschaert et al. (2008) concluded that these data indicate possible cross contamination from the slaughter equipment or transport crates.

8.4 *Salmonella* Contamination of Broiler Flocks in Different Countries in Europe

In 2008, a large report was produced by the European Food Safety Authority in the *EFSA Journal*. The research showed drastic differences in broiler flock prevalence for *Salmonella*. As found in the E.U. baseline study, the *Salmonella* broiler flock prevalence varied from 0% in Sweden, 1.6% in Denmark, 7.5% in the Netherlands, 15% in Germany, all the way up to 41.2% in Spain, 58.2% in Poland, and 68.2% in Hungary (European Food Safety Authority, 2008; Figure 8.2). These data indicate that, because the different countries within Europe use different methods for controlling *Salmonella* during grow out, flock prevalence varied significantly.

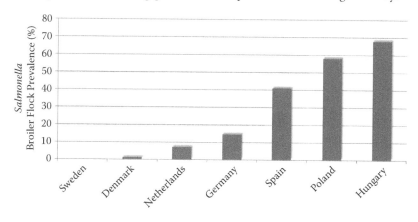

Figure 8.2 *Salmonella* **prevalence in broiler chicken flocks in different countries in Europe. (From European Food Safety Authority, 2008, A quantitative microbiological risk assessment on** *Salmonella* **in meat: Source attribution for human salmonellosis from meat. Scientific opinion of the panel,** *Journal of the European Food Safety Authority***, 625, 1–32.)**

8.5 Risk Factors for Having *Salmonella* in Flocks

Rosea et al. (1999) conducted studies to identify the risk factors for *Salmonella* contamination in French commercial broiler flocks at the end of the rearing period. A prospective study was carried out in 1996 and 1997 on 86 broiler flocks located in western France (Rosea et al., 1999). Litter was sampled for *Salmonella* using litter swabs and dust samples. Of the flocks, 70% had at least one contaminated environmental sample and were classified as *Salmonella*-contaminated flocks. The authors found that *Salmonella* contamination of the house before placing day-old chicks and the *Salmonella* contamination of day-old chicks were significantly related to *Salmonella* contamination of the flock at the end of the rearing period. The risk for *Salmonella* contamination of the flock increased if feed trucks were parked near the entrance of the house and if feed mash, instead of pellets, was provided to the chicks (Rosea et al., 1999).

Another research group conducted a similar study (Skov et al., 1999) to identify risk factors associated with *Salmonella enterica* serovar Typhimurium (*Salmonella* Typhimurium) infection in Danish broiler flocks. The data included all broiler flocks slaughtered in 1995. The authors reported that an increased risk for *Salmonella* Typhimurium infection was associated with two parent flocks, one confirmed infected and one suspected of being infected with *Salmonella* Typhimurium; with two of the hatcheries; and with five houses on the farm. An interaction between season and the previously mentioned hatcheries and a random effect at farm level was also statistically significant (Skov et al., 1999). Twelve variables evaluated in this study were not found to be associated with *Salmonella* Typhimurium infection: medication, growth promoters, breed of the laying flock, animal density, size of the flock, area of the house, age of the house, geographical location of the farm, observation of beetles, number of days between disinfection and replacement, visual appearance of the bedding, and age of the chickens when they were tested for *Salmonella* (Skov et al., 1999).

8.6 Effect of Free Range on *Salmonella* Prevalence

In 2007, Siemon et al. compared *Salmonella* prevalence between pasture-reared and conventionally reared broiler operations. Fecal droppings were collected from numerous farms around the United States over a 1-year period. At the farm level, authors detected a difference in *Salmonella* prevalence, with 33% of the flocks reared on pasture being positive and 47% of conventionally reared poultry flocks being positive (Siemon et al., 2007).

However, a new study by De Vylder et al. (2011) on egg layers reported the exact opposite. The researchers in Belgium suggested that any shift from conventional to alternative housing systems for laying hens should be accompanied by a keen concern for optimizing and maintaining *Salmonella* surveillance programs due to an increased risk of bird-to-bird transmission and internal egg contamination in some nontraditional housing systems. De Vylder et al. (2011) found that aviary and floor housing systems pose a greater risk of bird-to-bird transmission of *Salmonella* Enteritidis than traditional battery cages and furnished cages. Moreover, a higher number of eggs were internally contaminated by *Salmonella* in aviary systems, as compared to both cage systems and floor systems. The authors reported that the likely reasons for the difference in their findings and previous studies were that the previous findings may have been influenced by any number of factors, such as farm and flock size, or age of the housing system, that were not accounted for when interpreting the data.

8.7 Persistence of *Salmonella* in Soil

Salmonella has long been considered a ubiquitous organism. To be defined this way, a bacterium must be able to persist in soil for long periods of time. Soil may be a contamination route for *Salmonella* to enter poultry flocks. *Salmonella* may resist acidic soils for an extended length of time because the organism is able to mount an acid shock response. Moreover, pH has not been shown to affect the adhesion of *Salmonella* to soil particles (Stenström, 1989). Davies and Wray (1996a) isolated *Salmonella* Typhimurium from the soil surrounding calf burial pits 88 weeks after burial of contaminated carcasses.

Salmonella may also survive cleaning and disinfection procedures used on poultry farms. Pedersen et al. (2008) conducted a study to investigate the survival of *Salmonella senftenberg* in broiler breeder operations and broiler farms. Results indicated that *Salmonella senftenberg* had persisted for more than 2 years, despite cleaning, disinfection, desiccation, and depopulation, and was subsequently able to infect *Salmonella*-free breeders. This study demonstrated the difficulty in eliminating *Salmonella* from broiler breeder and broiler operations.

8.8 Epidemiological Tracking of Serotypes from Poultry to Humans

Gast (2007) stated that, of more than 2,500 identified *Salmonella* serotypes, only a small proportion are common in poultry flocks. However, there is an epidemiologically important connection between poultry products and human infections because many of the serotypes that are most prevalent in humans (such as *Salmonella* Typhimurium and *Salmonella* Enteritidis) are similarly common in poultry. Risk assessment studies have recommended intervention at multiple steps in the farm-to-table continuum as the most productive overall approach. Vaccination can enhance the short-term responsiveness of control programs to address problems involving specific serotypes of elevated significance (Gast, 2007).

8.9 How *Salmonella* Prevalence during Grow Out Can Influence *Salmonella* on the Finished Carcass

Russell (2007) reported that numerous factors during grow out can have a direct impact on the level of *Salmonella* on the finished product. The following is an example of how these factors, which have nothing to do with processing, can significantly influence carcass *Salmonella* results: In one instance, a company had two separate processing plants. One processing plant had a *Salmonella* prevalence of 17.5%. The other processing plant had a *Salmonella* prevalence of 6.7%. The company suspected that the plant that had 6.7% prevalence was doing an excellent job with processing, and that the plant with 17.5% *Salmonella* prevalence had a problem with flocks that were "hot" with *Salmonella*. When the sources of birds for the two plants were swapped with one another, instantaneously the *Salmonella* prevalence for the plant with the lower prevalence went from 6.7% to 25%. The *Salmonella* prevalence for the second plant fell from 17.5% to 1.6%. This effect held steady over a few months (Russell, 2007). These data clearly demonstrated that just by manipulating the flocks that go into a plant, the prevalence of *Salmonella* within that plant may be significantly impacted. Moreover, the final *Salmonella* prevalence may not be a good indicator of the ability of the plant to control *Salmonella* if excessive levels of *Salmonella* are introduced by processing heavily contaminated flocks.

8.10 Identification of Sources of *Salmonella* during Grow Out

8.10.1 Feed

A potential source of *Salmonella* contamination during grow out for poultry flocks that received great attention in the 1990s was poultry feedstuffs. As early as 1955, Erwin (1955) showed that *Salmonella* could be recovered from commercial poultry feed. Cox et al. (1983) studied mash and pelleted poultry feed and meat and bone meal samples to identify the bacterial species in them. *Salmonella* was found in 58%, 0%, and 92% of the mash, pelleted feeds, and meat and bone meal, respectively. These data show that the process of pelleting is effective for eliminating *Salmonella* from feed.

8.10.2 Decreasing *Salmonella* in Feed

Salmonella Enteriditis in poultry feed declines with increasing time of exposure to heat (Himathongkham et al., 1996). The interactions of temperature, time, and moisture and their effect on the thermal death of *Salmonella* Enteriditis were established in an experiment. The authors found that a linear relationship was obtained when the logarithm of survivors was plotted against the logarithm of exposure time of the feed to heat (Himathongkham et al., 1996).

In 1996, Davies and Wray (1996a) researched the survival of *Salmonella* Enteritidis in poultry food over a 2-year period. *Salmonella* persisted in poultry food troughs, and *Salmonella* Enteritidis survived at least 26 months in artificially contaminated poultry food.

The microbial decontamination of chicken feed using a direct-fired steam conditioner in a commercial pellet mill was evaluated (Blank et al., 1996). The prevalence of *Salmonella* in the feeds before conditioning was 8.3%. Following conditioning, *Salmonella* prevalence dropped to 1.7% (Blank et al., 1996). The authors found that when compared with a conventional, indirect-fired boiler-generated-steam conditioner (IFSC), the direct-fired steam conditioner proved superior with regard to pathogen decontamination.

8.10.3 *Effect of Cross Contamination of Feed*

The process of making poultry feed may contribute to *Salmonella* in feed due to cross contamination. A survey by Davies and Wray (1996a) detected *Salmonella* in 16% of poultry feed samples from a feed mill in England, with 24%, 13%, and 12% of samples from intake pits, ingredient bins, and mixers, respectively, contaminated. In 1994, Jones and Ricke also suggested that feed may be cross contaminated at the feed mill, and the authors suggested cleaning intake pits. Other research has shown that the source of the air used in making pellets is critical and should never originate from sites such as the ingredient receiving and loading areas (Jones and Ricke, 1994). The authors recommended having air ducts in sites that are protected from dust and other types of contamination (Jones and Ricke, 1994).

The presence of *Salmonella* in various poultry feeds has been well documented, and the frequency is influenced by various factors. For *Salmonella* to survive in feed, it must combat the same environmental conditions as nonpathogenic feed microflora. The most important bacterial inhibitor in poultry feed is low water activity. However, this factor alone is not able to completely destroy *Salmonella*. Research has shown that *Salmonella* has been isolated from poultry feed stored at 25°C after extended storage (16 months; Williams and Benson, 1978). Interestingly, Juven et al. (1984) reported that *Salmonella* survived better at a lower water activity of 0.43 than at a higher water activity level of 0.75. Survival and heat resistance of *Salmonella* spp. in meat and bonemeal, dry milk, and poultry feed are apparently inversely related to moisture content and relative humidity (Carlson and Snoeyenbos, 1970; Juven et al., 1984).

Many different *Salmonella* serotypes have been identified from poultry feed ingredients. *Salmonella* has been isolated from grain, oilseed meal, feather and fish meal, and meat and bonemeal. The only feed ingredient that is apparently resistant to contamination by *Salmonella* spp. is liquid animal and vegetable fat (Harris et al., 1997). Certain fatty acids have been shown to inhibit the growth of

Gram-negative bacteria (Khan and Katamy, 1969; Fay and Farias, 1975). Thus, it is important for feed producers to implement interventions to reduce or eliminate *Salmonella* in poultry feed.

8.11 Methods to Limit Salmonellae in Feeds: Chemical Amendments

Because mash types of poultry feed may contain *Salmonella,* feed producers have investigated numerous methods to reduce the levels of this pathogen in feed, including chemicals, heat, and irradiation. Jaquette et al. (1996) reported that chlorine was not effective, and that 2 mg/mL of chlorine was necessary to eliminate *Salmonella stanley* from alfalfa seeds. Compounds such as formic, hydrochloric, nitric, phosphoric, propionic, and sulfuric acids; isopropyl alcohol; formate and propionate salts; and trisodium phosphate have all been researched to determine their antibacterial properties in feed. Ha et al. (2000) reviewed some of the methods used to control *Salmonella* in feed as well as the specific application of chemical disinfectants to feed. The major concern when using chemicals in feed is that the chemical must perform well in the presence of high organic loads. Oxidant types of chemicals have difficulty performing well under these conditions. Most chemicals applied to feed would be applied during mixing; therefore, they must be proven not to be corrosive to mixing equipment. Another important consideration is that the chemicals should not decrease bird growth rates. Pritzl and Kienholz (1973) found that hydrochloric, sulfuric, phosphoric, and nitric acids added to feed at greater than 0.1 M will decrease the growth of the birds. Buffered organic acids, which are a natural and nontoxic component of intestinal digesta, have been widely used in poultry feed (Izat et al., 1990).

8.12 Methods to Limit Salmonellae in Feeds: Heat Treatment

Numerous scientists have researched the use of heat or pelleting for eliminating *Salmonella* from feed. Treatments such as cooking and pelleting usually involve heating the feed to temperatures between 70°C and 90°C (Halls and Tallentire, 1978; Furuta et al., 1980a, 1980b; McCapes et al., 1989). Williams (1981) summarized a series of results indicating that the effectiveness of particular heat treatments for limiting salmonellae in feed varied on the basis of time and temperature. The effectiveness of heat treatment may also be dependent on moisture content. Himathongkham et al. (1996) detected a 4.5 $\log_{10}$ reduction in *Salmonella* Enteritidis in poultry feed containing 15% moisture and held at 82.2°C for 2.2 seconds, but only a 1.5 $\log_{10}$ reduction in a similar treatment of feed containing 5% moisture. In addition, Matlho et al. (1997) found that heat treatments may be more effective if used in conjunction with chemical treatments such as propionic acid.

To eliminate *Salmonella* in protein meals such as meat and bone meal, cooking temperatures must be carefully monitored. Rasmussen et al. (1964) noted that cooking feed ingredients at 77°C for 15 minutes was not sufficient to consistently kill *Salmonella* spp. in meat and bone meal, but cooking at 82°C for 7 minutes was sufficient to destroy *Salmonella* to undetectable levels. Liu et al. (1969) reported that *Salmonella senftenberg* is more resistant to heat at 15% moisture than a feed with greater than 15% moisture and suggested heating regimens of 88°C and 15% moisture.

The prevalence of *Salmonella* in a dedicated commercial poultry feed mill was undertaken by Whyte et al. in 2003. The authors evaluated preheated samples and found that 18.8% of the feed contained *Salmonella*. After heat treatment, 22.6% of the feed contained *Salmonella*. Whyte et al. reported that feed ingredients and dust collected in the preheat treatment locations within the mill were frequently contaminated with *Salmonella* (11.8% and 33.3% of samples, respectively). High prevalences of *Salmonella* were also detected in dust samples (24.2%) obtained from the postheat treatment area of the mill and from feed delivery vehicles (57.1%). These results suggest that poultry feed became contaminated after heat treatments were applied in this feed mill.

Jones and Richardson (2004) collected feed ingredients, dust, and feed samples from three separate feed mills. Each of these mills produced between 100,000 and 400,000 tons of feed a year. The results clearly showed that feed ingredients and dust can be a major source of *Salmonella* contamination in feed mills. When samples were collected at the pellet mill, the authors noted an uneven distribution of *Salmonella* in feed as well as the need to control dust around the pellet mill (Jones and Richardson, 2004).

8.13 Effect of Contamination of Feed on Colonization of Birds

A group of researchers wanted to evaluate the impact of *Salmonella* contamination of feed on the colonization of baby chicks (Schleifer et al., 1984). The authors observed that only 1 *Salmonella* cell per gram of feed was required to colonize 1- to 7-day-old chicks. In a contradictory study with 8 million broiler chickens, Goren et al. (1988) reported that serotypes of *Salmonella* found on processed carcasses were related to those found in the hatchery but had no relation to serotypes found in feed samples. Most *Salmonella* are tolerant to desiccation and remain able to colonize and persist in feed mills (Pedersen et al., 2008).

In 2006, Huang et al. investigated the effects of feed particle size (coarse and fine) and feed form (mash and pellet) on the survival of *Salmonella* Typhimurium. These authors showed that pelleted poultry diets increased the prevalence of *Salmonella* Typhimurium in gizzards and ceca in growing broilers.

8.14 Influence of Insects

In a study by Kopanic et al. (1994), American cockroaches captured from a commercial poultry feed mill and hatchery were assayed for salmonellae. Five of 45 (11%) feed mill and 8 of 45 (17.8%) hatchery cockroach samples were confirmed positive for salmonellae. The authors suggested that cockroaches are capable of acquiring and infecting other cockroaches and objects, therefore implicating them as potential vectors of food-borne pathogens in poultry production and processing facilities.

In a study on the lesser mealworm (darkling beetle), Roche et al. (2006) determined if they could transmit marker *Salmonella* Typhimurium to day-of-hatch broiler chicks and potentially spread the organisms to nonchallenged penmates. Chicks were gavaged with beetles or larvae that were contaminated with a marker strain of *Salmonella* or gavaged with a saline solution containing the *Salmonella*. For the *Salmonella*-saline pens, 29–33% of the broilers that had been challenged and 10–55% of the penmates were positive at 3 weeks of age, and only 2–6% had positive ceca at 6 weeks. For the pens challenged with adult beetles, 0–57% of the challenged broilers and 20–40% of the penmates had positive ceca at 3 weeks, and 4–7% were positive at 6 weeks. The pens challenged with larvae had the greatest percentage of marker *Salmonella*-positive broilers; 25–33% of the challenged broilers and 45–58% of penmates were positive at 3 weeks, and 11–27% were positive at 6 weeks. The authors found that ingestion of larval or adult beetles contaminated with a marker *Salmonella* could be a significant vector for transmission to broilers.

8.15 Epidemiological Survey by the U.S. Department of Agriculture

In 2001, Bailey et al. with the Agricultural Research Service (ARS) of the U.S. Department of Agriculture (USDA) conducted a large epidemiological study to determine the relative importance of all known sources of *Salmonella* from the hatchery through grow out and processing in high- and low-production flocks from four integrated operations located in four states across four seasons (Bailey et al., 2001). The data obtained in this study are presented in Table 8.1.

Salmonella was recovered from most sample types throughout the entire grow-out period. The highest recovery rates were found in feces, litter, and drag swabs, indicating that the fecal oral horizontal transmission is critical in spreading *Salmonella* throughout the grow-out operation. Insects were found to play an important role as well. Bailey et al. (2001) reported that it was difficult to ascertain whether *Salmonella* was transferred from the insects and mice to the chickens or vice versa. However, the authors found that flies were highly contaminated (18.7%) and may play an important role in spreading *Salmonella*. Also, 6% of dirt samples

Table 8.1 *Salmonella* **Detection Percentage from All Sample Types, Times, and Integrators across All Seasons and High- and Low-Production Houses (32 Houses and 8,739 Samples)**

Sample	Total	Integrator A	Integrator B	Integrator C	Integrator D
Paper pads	50.8	32.5	47.4	26.6	96.0
Feces	6.6	0.8	10.3	9.2	5.2
Water line	1.4	0.0	1.3	0.0	3.7
Water cup	1.9	0.5	1.3	2.7	3.1
Litter	10.5	1.6	15.4	15.0	9.4
Feed hopper	2.3	1.6	3.9	0.0	3.2
Feeder	2.3	0.0	3.9	0.0	4.8
Drag swab	14.2	2.1	21.1	16.7	15.6
Wall swab	3.4	3.1	2.6	7.8	0.0
Fan swab	3.4	1.6	1.3	7.8	3.1
Mouse samples	6.1	12.5	0.0	3.7	0.0
Wild bird feces	6.6	6.1	14.3	4.3	4.8
Animal feces	3.0	3.8	0.0	0.0	2.6
Insects	2.8	6.1	0.8	4.2	1.9
Dirt, near entrance	6.1	6.3	13.5	3.3	0.0
Standing water	5.1	4.8	4.2	8.3	4.5
Boot swab	12.0	14.3	16.7	11.8	6.5
Fly strip	18.7	25.0	5.3	29.6	17.1
Cecal droppings	4.4	1.0	9.2	5.0	1.1
Total	9.8	5.2	10.8	9.7	13.4

Source: Bailey J S, Stern N J, Fedorka-Cray P, Craven S E, Cox N A, Cosby D E, Ladely S, and Musgrove M T, 2001, *Sources* and movement of *Salmonella* through integrated poultry operations: A multistate epidemiological investigation, *Journal of Food Protection*, 64, 1690–1697.

at the doors were contaminated with *Salmonella*, indicating that sanitizing boot dips should be used for farm workers entering and exiting chicken houses (Bailey et al., 2001). Henzler and Opitz (1999) found that mice were responsible for cross contamination of *Salmonella* in layer flocks. However, Bailey et al. (2001) found that only 8.3% of mouse rinses and 4% of mouse entrails were contaminated with *Salmonella*, indicating that these companies were controlling the rodent populations fairly well.

8.16 Effect of Season

Bailey et al. (2001) also presented data on the prevalence of *Salmonella* during different seasons of the year (Figure 8.3). In this study, *Salmonella* was highest in the fall and the lowest in the summer. It should be noted that this study was conducted across many different regions of the United States, and the spikes for *Salmonella* during various seasons should be different from region to region based on outside temperature, humidity levels, and management practices, such as use of vaccines by specific companies.

8.17 Effect of Feed Withdrawal

The modern broiler chicken has been bred over the years to be a veritable "eating machine." During grow out, broiler chickens eat approximately every 4 hours. Frequent eating is advantageous because birds that eat this frequently gain weight and put on edible muscle rapidly. This attribute may be considered a disadvantage

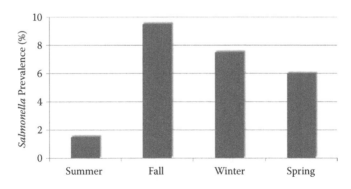

Figure 8.3 Prevalence of preharvest *Salmonella* **by season for all sample types and locations. (From Bailey J S, Stern N J, Fedorka-Cray P, Craven S E, Cox N A, Cosby D E, Ladely S, and Musgrove M T, 2001, Sources and movement of** *Salmonella* **through integrated poultry operations: A multistate epidemiological investigation,** *Journal of Food Protection*, **64, 1690–1697.)**

for maintaining the sanitary quality of the bird during processing. At the end of the grow-out period, prior to catching the birds and cooping them for transportation to the processing plant, the feed is removed from the birds for a period of approximately 3 to 7 hours. During this time, birds become hungry and begin to search for food. Because there is no food available to them in the feeders, they begin to search for feed on the floor, which may be contaminated. Barnhart et al. (1999a) reported that *Salmonella* contamination of the chicken crop has been reported to increase markedly and significantly during feed withdrawal, probably due to coprophagy (consumption of fecal material). This activity has been demonstrated to contribute significantly to the level of *Salmonella* on processed carcasses (Byrd et al., 2001). Studies have shown that many birds entering the processing plant have high levels of *Salmonella* in their crops as a result of this litter pecking (Byrd et al., 2001).

Salmonella in the crops of chickens that have consumed litter may be spread from carcass to carcass during the crop removal process (Hargis et al., 1995; Barnhart et al., 1999b). During cropping, the cropper piston is inserted into the vent area of the carcass and continues through the entire carcass, spinning as it goes. The piston has sharp grooves on the end of it that pick up the crop and that wrap the crop around the end of the cropper piston. As the piston moves through the neck opening, the cropper piston comes in contact with a brush that removes the crop from the piston. Then, the piston, while spinning, goes back through the entire carcass. If the crop breaks during this removal process, the contents leak onto the cropper piston and are transferred to the interior and exterior of the carcass, possibly spreading *Salmonella*.

8.17.1 Influence of the Crop

Studies have been conducted by Dr. Allen Byrd of the USDA-ARS in which the crops of live birds were filled with fluorescein dye. After 30 minutes, the birds were processed. By examining the carcasses at different stages of processing under a black light, crop contents that were transferred to the inside or outside of the carcass could be clearly visualized (S. M. Russell, unpublished data, 1998). These studies have shown that commercial croppers result in a large amount of contamination of the inside and outside of the carcasses. Thus, efforts should be made to control *Salmonella* in the crop prior to the crop removal process.

Some companies have been successful at controlling *Salmonella* in the crop by acidifying the bird's drinking water during the feed withdrawal process. Acetic, citric, or lactic acids and poultry water treatment (PWT) have all been used to acidify the crop to the extent that *Salmonella* are unable to survive. Byrd et al. (2001) found that lactic acid was most effective, and that 0.44% lactic acid in the waterers of broilers during the feed withdrawal period reduced *Salmonella*-contaminated crops by 80%. This effect carried over to the prechill carcasses, on which the prevalence of *Salmonella* was reduced by 52.4% (Byrd et al., 2001). When acidifying drinking water using lactic acid, it is best to expose the birds gradually to higher and higher levels of acid in the water the week before birds are to be caught. The key is to make the lactic acid concentration as high as possible while ensuring that the birds continue drinking the water. Suggestions for elimination of *Salmonella* in the crop prior to processing are as follows:

1. Apply an acidifier to drinking water of the chickens before the feed withdrawal period. Examples would be acetic acid, lactic acid, citric acid, or PWT.
2. Begin by applying small amounts of acid and gradually increase levels, making sure the birds are still consuming water and not backing away from the waterers.
3. Occasionally have the quality assurance employees check the pH of the crops of birds at the plant to ensure that they are being acidified.

8.17.2 Persistence of Salmonella in the Crops and Ceca

Hargis et al. (1995) evaluated the persistence of experimentally inoculated *Salmonella* Enteritidis in the crops and ceca of commercial broiler chickens during the last week of growth (Weeks 6 to 7) and the presence of crop and cecal *Salmonella* in 7-week-old broilers in a commercial processing plant. When broilers were inoculated with 10^6 colony-forming units (CFU) of *Salmonella* Enteritidis at 6 weeks of age by oral gavage, the incidence of crop and cecal contamination was equivalent 2 days after challenge (30%), with only 1 of 29 crops contaminated and of 29 ceca contaminated at 7 days following challenge. When broilers were inoculated with 10^8 CFU *Salmonella* Enteritidis at 6 weeks of age by oral gavage, 2 days after challenge, the crops and ceca were observed to be 57% and 67% positive for *Salmonella* Enteritidis, respectively. Seven days after inoculation with 10^8 CFU of *Salmonella* Enteritidis, the crops and ceca were 37% and 57% positive, respectively, for the challenge organism (Hargis et al. 1995). At a commercial broiler-processing plant, 286 of 550 crops were *Salmonella* positive, whereas only 73 of 500 ceca from these flocks were contaminated. The authors concluded that data from this plant indicated that the crops were far more likely to rupture than ceca (86-fold) during processing, increasing the possibility of carcass contamination with *Salmonella* derived from crop contents. The results of these studies suggest that the crop may serve as a significant source of carcass contamination with *Salmonella* within some processing plants.

Other researchers reported that crop contamination increased during preslaughter feed withdrawal and that contaminated crop contents may serve as an important source of *Salmonella* entry into poultry-processing plants (Corrier et al., 1999). The authors evaluated the effect of preslaughter feed withdrawal on crop pH and *Salmonella* crop contamination in broilers from three commercial broiler flocks. The effect of experimental feed withdrawal on crop pH, lactic acid concentration, and *Salmonella* crop contamination was also evaluated in market-age broilers challenged experimentally with *Salmonella* Typhimurium. Crop pH increased from 3.64 before feed removal to 5.14 after 8 hours of feed withdrawal in broilers from commercial flocks. Prevalence of *Salmonella* in the crops of the birds increased from 3.3% before feed removal to 12.6% after 8 hours of feed withdrawal (Hargis et al., 1995). *Salmonella* Typhimurium in the crops of the experimentally challenged broilers increased by approximately 1 $\log_{10}$ during the 8-hour experimental feed withdrawal. The authors concluded that feed withdrawal resulted in a decrease in lactic acid in the crop, accompanied by an increase in crop pH, and an increase in *Salmonella* crop contamination.

Ramirez et al. (1997) orally challenged broiler chickens with 10^8 CFU *Salmonella* Enteritidis at 6 weeks of age. At 7 weeks of age, the birds were randomly divided into two groups consisting of full access to feed or total feed withdrawal 18 hours prior to sample collection. At the time of sample collection, crops and ceca were aseptically removed and cultured for the presence or absence of *Salmonella* Enteritidis by enrichment (Ramirez et al., 1997). The prevalence of *Salmonella* Enteritidis-positive crops was consistently higher (2.8- to 7.3-fold increases) following feed withdrawal than the prevalence in samples collected from full-fed broilers. Similarly, the prevalence of *Salmonella* Enteritidis isolation was consistently higher (1.4- to 2.1-fold increases) in ceca following feed withdrawal than in samples collected from full-fed broilers.

In another experiment, ceca and crops were aseptically collected and cultured for the presence of *Salmonella* immediately prior to or following 8 hours of feed withdrawal at a commercial broiler house (Ramirez et al., 1997). Similar to the laboratory experiments, *Salmonella* were isolated more frequently from crops following feed withdrawal (36%) than from samples obtained immediately prior to withdrawal (19%). The authors reported that feed withdrawal increased the prevalence of *Salmonella* in broiler crops prior to slaughter and provided further evidence that the crop may be an important critical control point for reducing *Salmonella* contamination of broiler carcasses (Ramirez et al., 1997).

Crops or gizzards in broiler carcasses are frequently damaged during processing (Smith and Berrang, 2006). The contents from either organ, defined by the USDA Food Safety and Inspection Service as ingesta, may contain *Salmonella* and may contaminate the carcass. Previous research has shown crop contents are a source of *Salmonella* contamination on processed carcasses, although less information is available on gizzard contents. Smith and Berrang (2006) determined the prevalence and numbers of total aerobic bacteria, coliforms, *Escherichia coli*, and *Campylobacter* in ingesta collected from the crop and gizzard. Crop contents (log CFU/mL), compared with gizzard contents ($\log_{10}$ CFU/g), contained significantly ($p < 0.05$) higher numbers of total aerobic bacteria (5.6 vs. 2.9), coliforms (4.2 vs. 2.3), *E. coli* (3.9 vs. 2.2), and *Campylobacter* (4.6 vs. 2.2). *Escherichia coli* prevalence was higher in crop samples (28 of 29) than gizzard samples (19 of 30). *Campylobacter* prevalence was also higher for crop versus gizzard samples (29 of 29 vs. 12 of 30).

An average of 2.4 g of crop contents and 8.4 g of gizzard contents were recovered. Overall, the authors found that crop contents contained more bacteria than gizzard contents and contained a higher incidence of *E. coli* and *Campylobacter* contamination. However, because of the numbers of bacteria and amount of material in the crop and gizzard, it is unlikely that ingesta contamination would increase overall bacterial counts of prechill broiler carcasses.

Hinton et al. (2000b) conducted trials to determine the effect of feed withdrawal on the persistence of *Salmonella* Typhimurium in the ceca of market-age broiler chickens. Broilers were provided medicated or unmedicated feed and then were subjected to feed withdrawal for 0 to 24 hours in transportation crates or on litter. After withdrawing feed, broilers were stunned, bled, scalded, and picked, and the ceca were removed and evaluated. The number of total aerobes, Enterobacteriaceae, *Salmonella* Typhimurium, and lactic acid bacteria in the suspension were enumerated. Results indicated that there were significant increases in the population of Enterobacteriaceae during feed withdrawal (Hinton et al., 2000b). Feed withdrawal produced significant decreases in the population of lactic acid bacteria in all trials, but no significant change in the population of *Salmonella* Typhimurium occurred during feed withdrawal. There were no significant differences in *Salmonella* Typhimurium populations between broilers that were subjected to feed withdrawal on litter or in crates. The authors concluded that feed withdrawal did not always effectively evacuate the contents of the ceca, and that the ceca of broilers subjected to feed withdrawal can remain a source of food-borne bacterial pathogens; however, in this study, *Salmonella* populations did not increase as a result of feed withdrawal (Hinton et al., 2000b).

Consistent with the work of Hinton et al. (2000a), Rostagno et al. (2006) found that there were no significant effects of preslaughter practices on *Salmonella* prevalence in market-age turkeys. The authors stated that the difference found between earlier reports on broiler chickens and those found on turkeys may be based on the age of the birds. Broiler chickens are subjected to the preslaughter practices at approximately 6 to 8 weeks of age, whereas tom turkeys do not reach market age until they are approximately 18 to 21 weeks old (Rostagno et al., 2006). A possible reason why age of the bird may play a role is explained in studies that have shown the complexity of the intestinal bacterial community, as well as the immune response against *Salmonella* infections in broiler chickens, increases with age (Van der Wielen et al., 2002; Liu et al., 2003; Beal et al., 2005).

Rostagno et al. (2006) hypothesized that an established and more complex bacterial community, as well as a higher level of immune resistance in the intestinal tract of older market-age turkeys, may create a hostile environment for the establishment of new pathogens, such as *Salmonella*, particularly when birds are exposed to low doses of the bacteria. An established, complex intestinal microbial community and a mature immune system could prevent a significant increase in the prevalence of *Salmonella* due to new infections, even during exposure of market-age turkeys to a set of stressors associated with the preslaughter practices of the poultry industry.

Ramirez et al. (1997) conducted studies that demonstrated that feed withdrawal increases the incidence of *Salmonella* in broiler crops prior to slaughter and provides further evidence that the crop may be an important critical control point for reducing *Salmonella* contamination of broiler carcasses. Corrier et al. (1999) found that the incidence of *Salmonella* crop contamination may increase as much as fivefold during preslaughter feed withdrawal and represent a critical preharvest control point in reducing *Salmonella* entry into the processing plant. All of these studies pointed to the importance of controlling *Salmonella* in the crop of the birds during feed withdrawal.

Northcutt et al. (2003) conducted a study to determine effects of bird age at slaughter, feed withdrawal, and transportation on levels of coliforms, *Campylobacter*, *Escherichia coli*, and *Salmonella* on carcasses before immersion chilling. Broilers were processed at 42, 49, and 56 days of age after a 12-hour feed withdrawal period or a 0-hour feed withdrawal period (full fed). At each age, broilers were processed from two commercial farms previously identified as *Campylobacter* positive. One week before slaughter, broilers were gavaged with nalidixic acid-resistant *Salmonella*. Whole-carcass rinses (WCRs) were performed before chilling, and rinses were analyzed for coliforms, *Campylobacter*, *E. coli*, and *Salmonella*. Log$_{10}$ counts for coliforms, *Campylobacter*, and *E. coli* were affected by bird age at slaughter. Feed withdrawal affected only *Campylobacter* on carcasses of older broilers (56 days of age). Under the conditions of this experiment, it seemed that contamination on the exterior of birds entering the processing facility is critical to carcass bacterial counts. The authors reported that feed withdrawal may increase prechill carcass counts for *E. coli* and *Campylobacter*, but not *Salmonella*.

To reduce *Salmonella* and *Campylobacter* in the crop of birds, Byrd et al. (2001) evaluated the use of selected organic acids (0.5% acetic, lactic, or formic) in drinking water during a simulated 8-hour pretransport feed withdrawal. *Salmonella* Typhimurium was recovered from 53% of the control crops and from 45% of crops from acetic acid-treated broilers. However, treatment with lactic acid (31%) or formic acid (28%) reduced prevalence more than the acetic acid.

In another study on a commercial farm, broilers were provided 0.44% lactic acid during a 10-hour feed withdrawal period (Byrd et al., 2001). Prechill carcass rinse samples were collected for *Campylobacter* and *Salmonella* detection. Crop contamination with *Salmonella* was significantly reduced when lactic acid was used (3.4%) as compared with controls (16.6%). Importantly, *Salmonella* isolation prevalence in prechill carcass rinses was significantly reduced by 52.4% with the use of lactic acid (14.9% vs. 31.2%). Crop contamination with *Campylobacter* was significantly reduced by lactic acid treatment (62.3%) as compared with the controls (85.1%). Lactic acid also reduced the incidence of *Campylobacter* found on prechill carcass rinses by 14.7% compared with the controls. These studies suggested that incorporation of lactic acid in the drinking water during pretransport FW may reduce *Salmonella* and *Campylobacter* contamination of crops and broiler carcasses at processing.

Hinton et al. (2000b) challenged broiler chickens with 10^9 *Salmonella* Typhimurium and were then provided a glucose-based cocktail supplemented with 0–15% glucose during feed withdrawal in battery cages or in pens on litter. The authors found that fewer *Salmonella* Typhimurium and other Enterobacteriaceae were recovered from the crops of broilers provided the cocktail supplemented with 7.5% glucose than from the crops of broilers provided either water or cocktails supplemented with lower or higher concentrations of glucose. Inhibition of the growth of *Salmonella* Typhimurium and other Enterobacteriaceae in the crops of broilers provided the cocktail supplemented with 7.5% glucose was generally associated with increased growth of lactic acid bacteria and decreased crop pH. Hinton et al. (2000b) encouraged providing the cocktail to broilers before shipping to processing plants to reduce the number of food-borne pathogens that poultry carry into processing plants.

8.18 Effect of Transport Stress

The stress due to feed withdrawal and transportation may increase *Salmonella* prevalence in broiler chickens. Line et al. (1997) reported that the stresses associated with transporting poultry prior to slaughter have been shown to increase pathogen populations both in the intestinal tract and on the carcass exterior.

References

Bailey J S, Stern N J, Fedorka-Cray P, Craven S E, Cox N A, Cosby D E, Ladely S, and Musgrove M T (2001), Sources and movement of *Salmonella* through integrated poultry operations: A multistate epidemiological investigation, *Journal of Food Protection*, 64, 1690–1697.

Barnhart E T, Caldwell D J, Crouch M C, Byrd J A, Corrier D E, and Hargis B M (1999a), Effect of lactose administration in drinking water prior to and during feed withdrawal on *Salmonella* recovery from broiler crops and ceca, *Poultry Science*, 78, 211–214.

Barnhart E T, Sarlin L L, Caldwell D J, Byrd J A, Corrier D E, and Hargis B M (1999b), Evaluation of potential disinfectants for preslaughter broiler crop decontamination, *Poultry Science*, 78, 32–37.

Beal R K, Powers C, Wigley P, Barrow P A, Kaiser P, and Smith A L (2005), A strong antigen-specific T-cell response is associated with age and genetically dependent resistance to avian enteric salmonellosis, *Infection and Immunity*, 73, 7509–7516.

Blank G, Savoie S, and Campbell L D (1996), Microbiological decontamination of poultry feed-evaluation of steam conditioners, *Journal of the Science of Food and Agriculture*, 72, 299–305.

Byrd J A, Anderson R C, Brewer R L, Callaway T R, Bischoff K M, Mcreynolds J L, Caldwell D J, Hargis B M, Herron K L, and Bailey R H (2001), Effect of lactic acid administration in the drinking water during preslaughter feed withdrawal on *Salmonella* and *Campylobacter* contamination of broilers, *Poultry Science*, 80, 278–283.

Carlson V L, and Snoeyenbos H S (1970), Effect of moisture on *Salmonella* populations in animal feeds, *Poultry Science*, 49, 717–725.

Corrier D E, Byrd J A, Hargis B M, Hume M E, Bailey R H, and Stanker L H (1999), Presence of *Salmonella* in the crop and ceca of broiler chickens before and after pre-slaughter feed withdrawal, *Poultry Science*, 78, 45–49.

Cox N A, Bailey J S, and Berrang M E (1996), Alternative routes for *Salmonella* intestinal tract colonization of chicks, *Journal of Applied Poultry Research*, 5, 282–288.

Cox N A, Bailey J S, Thomson J E, and Juven B J (1983), *Salmonella* and other Enterobacteriaceae found in commercial poultry feed, *Poultry Science*, 62, 2169–2175.

Davies R H, and Wray C (1996a), Persistence of *Salmonella* Enteritidis in poultry units and poultry food, *British Poultry Science*, 37, 589–596.

Davies R H, and Wray C (1996b), Studies of contamination of three broiler breeder houses with *Salmonella* Enteritidis before and after cleansing and disinfection, *Avian Diseases*, 40, 626–633.

De Vylder J, Dewulf J, Van Hoorebeke S, Pasmans F, Haesebrouck F, Ducatelle R, and Van Immerseel F (2011), Horizontal transmission of *Salmonella* Enteritidis in groups of experimentally infected laying hens housed in different housing systems, *Poultry Science*, 90, 1391–1396.

Erwin L E (1955), Examination of prepared poultry feeds for the presence of *Salmonella* and other enteric organisms, *Poultry Science*, 34, 215–216.

European Food Safety Authority (2008), A quantitative microbiological risk assessment on *Salmonella* in meat: Source attribution for human salmonellosis from meat. Scientific opinion of the panel, *Journal of the European Food Safety Authority*, 625, 1–32.

Fay J P, and Farias R N (1975), The inhibitory action of fatty acids on the growth of *Escherichia coli*, *Journal of General Microbiology*, 91, 233–240.

Furuta K, Morimoto S, and Sato S (1980a), Bacterial contamination in feed ingredients, for-mulated chicken feed and reduction of viable bacteria by pelleting, *Laboratory Animals*, 14, 221–224.

Furuta K, Oku I, and Morimoto S (1980b), Effect of steam temperature in the pelleting process of chicken food on the viability of contaminating bacteria, *Laboratory Animals*, 14, 293–296.

Gast R K (2007), Serotype-specific and serotype-independent strategies for preharvest con-trol of food-borne *Salmonella* in poultry, *Avian Diseases*, 51, 817–828.

Goren E, Dejong W A, Doornebal P, Bolder N M, Mulder R W A W, and Jansen A (1988), Reduction of salmonellae infection of broilers by spray application of intestinal micro-flora: a longitudinal study, *Veterinary Quarterly*, 10, 249–255.

Ha S D, Maciorowski K G, and Ricke S C (2000), Application of antimicrobial approaches for reducing *Salmonella* contamination in poultry feed: A review, *Research Advances in Antimicrobial Agents and Chemotherapy*, 1, 19–33.

Halls N A, and Tallentire A (1978), Effects of processing and gamma irradiation on the micro-biological contaminants of a laboratory animal diet, *Laboratory Animals*, 12, 5–10.

Hargis B M, Caldwell D J, Brewer R L, Corrier D E, and Deloach J R (1995), Evaluation of the chicken crop as a source of *Salmonella* contamination for broiler carcasses, *Poultry Science*, 74, 1548–1552.

Harris I T, Fedorka-Cray P J, Gray J T, Thomas L A, and Ferris K (1997), Prevalence of *Salmonella* organisms in swine feed, *Journal of Applied Veterinary Medicine Association*, 210, 382–385.

Henzler D J, and Opitz J M (1999), Role of rodents in the epidemiology of *Salmonella* enterica serovar Enteritidis and other *Salmonella* serovars in poultry farms, in Salmonella *enterica serovar Enteritidis in Humans and Animals*, ed A M Saeed, Iowa State University Press, Ames, IA, 331–340.

Himathongkham S, Das M, Pereira G, and Riemann H (1996), Heat destruction of *Salmonella* in poultry feed: Effect of time, temperature, and moisture, *Avian Diseases*, 40, 72–77.

Hinton A, Jr, Buhr R J, and Ingram K D (2000a), Physical, chemical, and microbiological changes in the ceca of broiler chickens subjected to incremental feed withdrawal, *Poultry Science*, 79, 483–488.

Hinton A, Jr, Buhr R J, and Ingram K D (2000b), Reduction of *Salmonella* in the crop of broiler chickens subjected to feed withdrawal, *Poultry Science*, 79, 1566–1570.

Huang D S, Li D F, Xing J J, Ma Y X, Li Z J, and Lv S Q (2006), Effects of feed particle size and feed form on survival of *Salmonella* Typhimurium in the alimentary tract and cecal *Salmonella* Typhimurium reduction in growing broilers, *Poultry Science*, 85, 831–836.

Izat A L, Tidwell N M, Thomas R A, Reiber M A, Adams M H, Colberg M, and Waldroup P W (1990), Effect of a buffered propionic acid in diets on the performance of broiler chickens and on microflora of the intestine and carcass, *Poultry Science*, 69, 818–826.

Jaquette C B, Beuchat L R, and Mahon B E (1996), Efficacy of chlorine and heat treatment in killing *Salmonella stanley* inoculated onto alfalfa seeds and growth and survival of the pathogen during sprouting and storage, *Applied Environmental Microbiology*, 62, 2212–2215.

Jones F, Axtell R C, Tarver F R, Rives D V, Scheideler S E, and Wineland M J (1991), Environmental factors contributing to *Salmonella* colonization of chickens, in *Colonization Control of Human Bacterial Enteropathogens in Poultry*, ed L C Blankenship, Academic Press, San Diego, CA, 3–20.

Jones F T, and Richardson K E (2004), *Salmonella* in commercially manufactured feeds, *Poultry Science*, 83, 384–391.

Jones F T, and Ricke S C (1994), Researchers propose tentative HACCP plan for feed mills, *Feedstuffs*, 66, 32, 36–42.

Juven B J, Cox N A, and Bailey J S (1984), Survival of *Salmonella* in dry food and feed, *Journal of Food Protection*, 47, 445–448.

Khan M, and Katamy M (1969), Antagonistic effect of fatty acids against *Salmonella* in meat and bone meal, *Applied Microbiology*, 17, 402–404.

Klowden M J, and Greenberg B (1976), *Salmonella* in the American cockroach: Evaluation of vector potential through dosed feeding experiments, *Journal of Hygiene*, 77, 105–111.

Klowden M J, and Greenberg B (1977), Effects of antibiotics on the survival of *Salmonella* in the American cockroach, *Journal of Hygiene*, 79, 339–345.

Kopanic R J, Jr, Sheldon B W, Wright C G (1994), Cockroaches as vectors of *Salmonella*: Laboratory and field trials, *Journal of Food Protection*, 57, 125–132.

Line J E, Bailey J S, Cox N A, and Stern N J (1997), Yeast treatment to reduce *Salmonella* and *Campylobacter* populations associated with broiler chickens subjected to transport stress, *Poultry Science*, 76, 1227–1231.

Liu T S, Snoeyenbos G H, and Carlson V L (1969), Thermal resistance of *Salmonella senftenberg* 775W in dry animal feeds, *Avian Diseases*, 13, 611–631.

Liu W, Kaiser M G, and Lamont S J (2003), Natural resistance-associated macrophage protein 1 gene polymorphisms and response to vaccine against or challenge with *Salmonella* Enteritidis in young chicks, *Poultry Science*, 82, 259–266.

Matlho G, Himathongkham S, Riemann H, and Kass P (1997), Destruction of *Salmonella* Enteritidis in poultry feed by combination of heat and propionic acid, *Avian Diseases*, 41, 58–61.

McCapes R H, Ekperigin H E, Cameron W J, Ritchie W L, Slagter J, Stangeland V, and Nagaraja K V (1989), Effect of a new pelleting process on the level of contamination of poultry mash by *Escherichia coli* and *Salmonella*, *Avian Diseases*, 33, 103–111.

Northcutt J K, Berrang M E, Dickens J A, Fletcher D L, and Cox N A (2003), Effect of broiler age, feed withdrawal, and transportation on levels of coliforms, *Campylobacter*, *Escherichia coli* and *Salmonella* on carcasses before and after immersion chilling, *Poultry Science*, 82, 169–173.

Pedersen T B, Bisgaard M, and Olsen J E (2008), Persistence of *Salmonella senftenberg* in poultry production environments and investigation of its resistance to desiccation, *Avian Pathology*, 37, 421–427.

Pritzl M C, and Kienholz E W (1973), The effect of hydrochloric, sulfuric, phosphoric, and nitric acids in diets for broiler chicks, *Poultry Science*, 52, 1979–1981.

Ramirez G A, Sarlin L L, Caldwell D J, Yezak C R, Hume M E, Corrier D E, Deloach J R, and Hargis B M (1997), Effect of feed withdrawal on the incidence of *Salmonella* in the crops and ceca of market age broiler chickens, *Poultry Science*, 76, 654–656.

Rasmussen O G, Hansen R, Jacobs N J, and Wilder O H M (1964), Dry heat resistance of salmonellae in rendered animal by-products, *Poultry Science*, 43, 1151–1157.

Rasschaert G, Houf K, Godard C, Wildemauwe C, Pastuszczak-Frak M, and De Zutter L (2008), Contamination of carcasses with *Salmonella* during poultry slaughter, *Journal of Food Protection*, 71, 146–152.

Roche A J, Cason J A, Fairchild B D, Hinkle N C, Cox N A, Richardson L J, and Buhr R J (2006), Transmission of *Salmonella* to broilers by contaminated larval and adult lesser mealworms, *Alphitobius diaperinus* (Coleoptera: Tenebrionidae), *Poultry Science*, 88, 44–48.

Rosea N, Beaudeaub F, Drouina P, Touxa J Y, Rosea V, and Colina P (1999), Risk factors for *Salmonella* enterica subsp. Enterica contamination in French broiler-chicken flocks at the end of the rearing period, *Preventative Veterinary Medicine*, 39, 265–277.

Rostagno M H, Wesley I V, Trampel D W, and Hurd H S (2006), *Salmonella* prevalence in market-age turkeys on-farm and at slaughter, *Poultry Science*, 85, 1838–1842.

Russell S M (2007), Pre-harvest *Salmonella* reduction: The next step in control, *Poultry USA Magazine*, 32, 34–35.

Schleifer J J, Juven B J, Beard C W, and Cox N A (1984), The susceptibility of chicks to *Salmonella montevideo* in artificially contaminated poultry feed, *Avian Diseases*, 28, 497–503.

Siemon C E, Bahnson P B, and Gebreyes W A (2007), Comparative investigation of prevalence and antimicrobial resistance of *Salmonella* between pasture and conventionally reared poultry, *Avian Diseases*, 51, 112–117.

Skov M N, Angen O, Chriel M O, Olsen J E, and Bisgaard M (1999), Risk factors associated with *Salmonella* enterica serovar Typhimurium infection in Danish broiler flocks, *Poultry Science*, 78, 848–854.

Smith D P, and Berrang M E (2006), Prevalence and numbers of bacteria in broiler crop and gizzard contents, *Poultry Science*, 85, 144–147.

Stenström T A (1989), Bacterial hydrophobicity, an overall parameter for the measurement of adhesion potential to soil particles, *Applied and Environmental Microbiology*, 55, 142–147.

Van Der Wielen P W J J, Keuzenkamp D A, Lipman L J A, Van Knapen F, and Biesterveld S (2002), Spatial and temporal variation of the intestinal bacterial community in commercially raised broiler chickens during growth, *Microbial Ecology*, 44, 286–293.

Whyte P, McGill K, and Collins J D (2003), A survey of the prevalence of Salmonella and other enteric pathogens in a commercial feed mill, *Journal of Food Safety*, 23, 13–24.

Williams J E (1981), Salmonellas in poultry feeds—A worldwide review. III. Methods in control and elimination, *Worlds Poultry Science Journal*, 37, 97–105.

Williams J E, and Benson S T (1978), Survival of *Salmonella typhimurium* in poultry feed and litter at three temperatures, *Avian Diseases*, 22, 742–747.

Chapter 9

Salmonella Intervention during Grow Out

9.1 Introduction

Many attempts have been used to decrease *Salmonella* in broiler chickens during grow out. The methods used vary greatly depending on the area of the world where the chickens are grown and to which area of the world the final product will be exported. The most commonly used methods are competitive exclusion (CE) and vaccination.

9.2 Competitive Exclusion

As long ago as 1973, scientists were considering feeding beneficial bacteria to baby chicks to prevent colonization and growth of *Salmonella* in their intestinal tract. Because baby chicks have no intestinal flora when hatched, it would be advantageous to establish mature, healthy gut microflora to chicks to prevent colonization of the chicks by *Salmonella*. Rantala and Nurmi (1973) were the original scientists who proposed this idea. These researchers inoculated 2-day-old chicks with *Salmonella infantis* after pretreating the chicks with the cultured microflora of the alimentary tract of an adult chicken as well as fluids from the alimentary tract. The authors demonstrated that these cultures, which came to be known as *competitive exclusion* (CE) cultures, prevented the colonization of the ceca by *Salmonella infantis*. For the control chicks that were not treated with the CE cultures, 10^7 *Salmonella infantis*/g were isolated from the ceca. These mature gut flora CE cultures were found to be effective for preventing chicks from being colonized by *Salmonella infantis*.

In 1982, Impey prevented colonization of the ceca of new chicks by *Salmonella* Typhimurium by administering a mixture of cultures comprising 48 different bacterial strains originating from an adult bird known to be free of *Salmonella*. In this study, the treatment conferred protection to the same degree as a suspension of adult cecal contents or an undefined anaerobic culture from the same source and was demonstrated in four separate laboratory trials (Impey, 1982). The presence of high levels of lactobacilli and *Bacteroides* spp., which are not found usually at 2 days of age in chicks produced under commercial conditions, was indicative of the successful establishment of an adult-type microflora.

Wierup et al. (1992) reported that, in Sweden, CE cultures have been used since 1981 as a part of the National Control Program for *Salmonella*. In this program, all broiler flocks are tested for *Salmonella* before slaughter, providing an evaluation of CE treatment. During the period 1981–1990, CE culture was given to 179 flocks, involving 3.82 million chickens. Only one of the treated flocks was found to be *Salmonella* positive. Wierup et al. (1992) stated that the control program has led to a very low incidence of *Salmonella* in broiler chickens in Sweden. A nationwide study carried out in 1990 demonstrated that less than 1% of broiler chickens were contaminated with *Salmonella* after slaughter.

Likewise, Nurmi et al. (1992) reported that Finland began widespread use of CE treatments for poultry, and in 1992, over 70% of growers were using it routinely. The number of *Salmonella*-positive flocks was less than 5%, and the incidence of *Salmonella*-contaminated broiler carcasses has been 5–11% in the 2 years prior to the study (Nurmi et al., 1992). The average number of *Salmonella* cells on contaminated carcasses was very low, with generally less than five cells per carcass. The authors stated that most (70–80%) of human *Salmonella* infections are contracted abroad (outside the Nordic countries). Nurmi et al. (1992) stated that only 15–20% of some 1,200 cases of domestic origin are caused by contaminated poultry.

9.3 Use of Sugars in Combination with CE Cultures

Corrier et al. (1993a) proposed the idea of administering lactose in combination with used poultry litter containing cecal and fecal droppings from adult broilers. The authors evaluated their protective effects against *Salmonella enteritidis* colonization in leghorn chicks and 16-week-old hens. The addition of used litter as 5% of the feed ration significantly decreased ($p < 0.01$) *Salmonella* cecal and organ colonization in the chicks. In this study, provision of used litter or used litter and lactose in the feed failed to provide protection against *Salmonella* colonization in the hens.

In a follow-up study, Corrier et al. (1993b) evaluated a defined mixed culture of indigenous cecal bacteria from adult broilers in terms of their protective effects on *Salmonella* Typhimurium colonization of broiler chicks. Compared with controls, the mean number of *Salmonella* Typhimurium in the cecal contents of the chicks given CE culture and dietary lactose decreased by 4.2 $\log_{10}$ CFU/g. Similarly, the numbers of *Salmonella* cecal culture-positive chicks was decreased by 55% in the chicks given CE culture and lactose (Corrier et al., 1993b). The authors concluded that these results indicated that a defined culture of indigenous cecal bacteria isolated and maintained in CE culture, together with dietary lactose, effectively controlled *Salmonella* Typhimurium cecal colonization in newly hatched broiler chicks.

In 1993, Blankenship et al. employed a two-step treatment of broiler chicks with a mucosal competitive exclusion (MCE) culture in which the MCE was first sprayed on chicks in the hatchery followed by administration in the first drinking water in commercial broiler flocks. The authors reported that initial feed, water, and litter contamination was at a low frequency (<10%), eggshell fragments and chick paper pads were frequently contaminated (>50%), and after 3 weeks of growth, contamination of litter, skin with feathers, and ceca was reduced in treated flocks as compared with control flocks. *Salmonella* prevalence in ceca and in processed carcass rinses was also reduced from 41% in control flocks to 10% in treated flocks (Figure 9.1).

A new approach to CE was taken by Hollister et al. in 1994. Cultures of cecal bacteria that were grown under anaerobic conditions were prepared as lyophilized powder or encapsulated and lyophilized in alginate beads and compared with broth cultures for control of *Salmonella* Typhimurium colonization and for ease of delivery. Cultures delivered by gavaging the birds and those encased in alginate beads with dietary lactose reduced mean *Salmonella* colony-forming units in cecal contents by 3.4 to 5.3 $\log_{10}$ CFU/g at 10 days of age (Hollister et al., 1994).

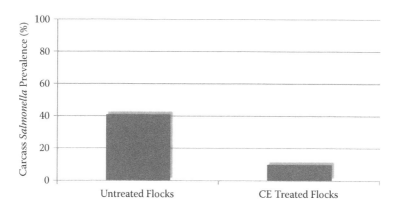

Figure 9.1 *Salmonella* **prevalence on carcasses from birds treated with CE versus untreated controls. (From Blankenship L C, Bailey J S, Cox N A, Stern N J, Brewer R, and Williams O, 1993, Two-step mucosal competitive exclusion flora treatment to diminish salmonellae in commercial broiler chickens,** *Poultry Science***, 72, 1667–1672.)**

Instead of using bacteria as the CE culture, Line et al. (1997) investigated the use of the yeast *Saccharomyces boulardii* for its ability to reduce populations of *Salmonella* and *Campylobacter* in broiler chickens subjected to transport stress. Chicks inoculated with individual strains of *Salmonella* and *Campylobacter* were held 6 weeks and then divided into two groups, with half of the chickens receiving 10% dried yeast in the feed for 60 hours. With no yeast treatment, transport stress caused the *Salmonella* colonization frequency to increase more than fivefold, from 3.3% to 16.7% (Line et al., 1997). No *Salmonella* were recovered from the yeast-treated birds. This study demonstrated the viability of using *Saccharomyces boulardii* as a means of preventing colonization of broiler chickens with *Salmonella*.

9.4 Effect of Feed Ingredients on Eliminating *Salmonella*

Many studies have been conducted seeking to add feed ingredients to poultry feed that will prevent the colonization of the birds with *Salmonella*. As early as 1991, Bailey et al. researched the use of fructooligosaccharide (FOS) on the ability of *Salmonella* Typhimurium to grow and colonize the intestinal tract of chickens. *Salmonella* were not able to multiply when FOS was the sole carbon source. When FOS was fed to chicks at 0.375%, little reduction of *Salmonella* was observed; however, at 0.75%, there were 12% fewer FOS-fed birds colonized with *Salmonella* when compared with control birds. When chicks given a partially protective CE culture were fed diets supplemented with 0.75% FOS, only 4 of 21 (19%) chickens challenged with 10^9 *Salmonella* cells on Day 7 became colonized as compared with 14 of 23 (61%) chickens given CE alone. Bailey et al. (1991) reported that chickens treated with FOS had a fourfold reduction in the level of *Salmonella* present in the ceca. The authors stated that feeding FOS in the diet of chickens may lead to a shift in the intestinal gut microflora and under some circumstances may result in reduced susceptibility to *Salmonella* colonization.

Cox et al. (1992) investigated broiler rations supplemented with 0.6% acid (either butyric or lactic) starting at 1 day of age, while a control group received unsupplemented feed. At 2 days of age, chicks were gavaged with 10^6 *Salmonella* Typhimurium. After 14 days, lactic acid decreased colonization by 1.6 logs, while no significant decrease was observed with butyric acid. After 21 days, both acids had significantly reduced intestinal colonization; butyric and lactic acid produced 1.67 and 1.95 $\log_{10}$ reductions, respectively (Cox et al., 1992).

In an attempt to enrich cecal bifidobacterial populations and reduce colonization levels of *Salmonella* in the ceca of broiler chickens, Thitaram et al. (2005) used a compound derived by fermentation called isomaltooligosaccharide (IMO). Broiler diets were prepared with final IMO concentrations of 1%, 2%, and 4%, and a control diet was prepared containing no IMO supplementation.

Chickens were challenged with 10^8 cells of *Salmonella enterica* ser. Typhimurium. The authors reported that IMO-supplemented diets resulted in higher cecal bifidobacteria compared with the control diet. However, there was no significant difference in bifidobacterial counts among the treatment groups. Chickens fed diets with 1% IMO had a 2 $\log_{10}$ reduction in the level of inoculated *Salmonella* Typhimurium present in the ceca compared with the control group (Thitaram et al., 2005).

The protective efficacy of activated charcoal containing wood vinegar liquid (Nekka-Rich) against intestinal infection with *Salmonella enterica* serovar Enteritidis (SE) was evaluated by Watarai and Tana (2005). The authors found that SE was effectively adsorbed by activated charcoal of Nekka-Rich. Chickens were fed a basal diet containing Nekka-Rich or immunized with commercially obtained SE vaccine and challenged with SE. Significantly less fecal excretion of SE was observed in chickens fed Nekka-Rich for 10 days after challenge. On Day 15 after challenge, the authors were not able to isolate SE from fecal samples. On the other hand, immunization of chickens with SE vaccine did not fully inhibit bacterial growth. Watarai and Tana (2005) reported that fecal excretion of SE was consistently observed in the vaccinated chickens after challenge. Nekka-Rich charcoal was effective for eliminating SE shedding in chickens.

9.5 Use of Bacteriophages to Kill *Salmonella* during Grow Out

A *bacteriophage* (BP) is defined as any of a number of viruses that infect and kill bacteria (Figure 9.2). BPs are among the most common biological entities on Earth (Collman, 2001). Phages are estimated to be the most widely distributed and diverse entities in the biosphere (McGrath and van Sinderen, 2007). BPs are extremely specific and can distinguish between species of bacteria. For example, a phage that only kills SE may be isolated and used to select and kill this organism. As such, scientists have attempted to use BPs to kill specific species of bacteria on live poultry and on processed carcasses.

Toro et al. (2005) tested *Salmonella*-specific BPs for their ability to reduce *Salmonella* colonization in experimentally infected chickens. A "cocktail" of distinct phage (i.e., phage showing different host ranges and inducing different types of plaques on *Salmonella* Typhimurium cultures) was developed. The authors reported a reduction in *Salmonella* counts in the ceca and ileal sections of BP-treated chickens as compared with nontreated birds. Overall, the results indicated a protective effect of *Salmonella*-specific BPs against *Salmonella* colonization of experimentally infected chickens (Toro et al., 2005).

Producers of free-range poultry have difficulty controlling *Salmonella* on fully processed products because they are limited to very few interventions. BPs isolated from free-range chickens were analyzed as a therapeutic agent for reducing the

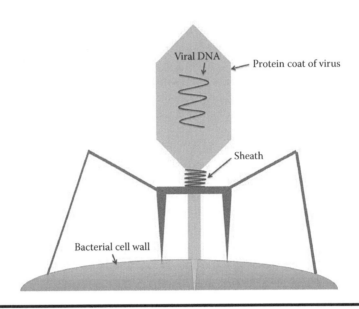

Figure 9.2 Diagram of a bacteriophage attacking a bacterial cell. (From http://www.salmonellablog.com/salmonella-watch/scientists-study-natures-toolbox-to-identify-and-destroy-salmonella/.)

concentration of *Salmonella enterica* serovar Enteritidis phage Type 4 (*Salmonella* Enteritidis PT4) in the ceca of broilers (Fiorentin et al., 2005). One-day-old broilers infected with *Salmonella* Enteritidis PT4 were orally treated on Day 7 with a mixture of 10^{11} plaque-forming units (PFU) of each of three BPs. In 5 days, the BP-treated group showed a reduction of 3.5 orders of magnitude on colony-forming units (CFU) of *Salmonella* Enteritidis PT4 per gram of cecal material. Even after 10, 15, 20, and 25 days, birds still had fewer CFUs of *Salmonella* Enteritidis PT4 per gram of cecal material (Fiorentin et al., 2005).

In 2007, Andreatti-Filho et al. used SE-lysing BPs isolated from poultry or human sewage sources to reduce SE in experimentally infected chicks. Cocktails of 4 different BPs obtained from commercial broiler houses and 45 BPs from a municipal wastewater treatment plant were evaluated. The phages significantly reduced SE recovery from cecal tonsils at 24 hours. In another experiment, day-of-hatch chicks were challenged orally with 9×10^3 CFU/chick SE and gavaged with 1×10^8 poultry BP PFU/chick or 1.2×10^8 human sewage BP PFU/chick, or a combination of both 1 hour postchallenge. All treatments reduced SE recovered from cecal tonsils at 24 hours as compared with untreated controls, but no significant differences were observed at 48 hours following treatment. These data suggest that some BPs can be efficacious in reducing SE colonization in poultry for a short period, but with the BPs and methods presently tested, persistent reductions were not observed (Andreatti-Filho et al., 2007). It is important to note that the bacteria become resistant to these phages, as seen in this study.

In 2008, Borie et al. isolated three different lytic BPs from the sewage system of commercial chicken flocks and used them to reduce SE colonization of experimental chickens. Ten-day-old chickens were challenged with 9.6×10^5 CFU/mL of a SE strain and treated by coarse spray or drinking water with a cocktail of the three phages at 10^3 PFU 24 hours prior to SE challenge. At Day 20, the chickens were euthanized, and the intestines were evaluated. Aerosol spray delivery of BPs significantly reduced the incidence of SE infection in the chicken group ($p = 0.0084$) to 72.7% as compared with the control group (100%). In addition, SE counts showed that phage delivery by both coarse spray and drinking water reduced the intestinal SE colonization (Borie et al., 2008). The authors found that phage treatment, by either aerosol spray or drinking water, is a plausible alternative to antibiotics for the reduction of *Salmonella* infection in poultry.

9.6 Vaccination

The use of vaccination for controlling *Salmonella* has been proposed for a long time. As early as 1971, Knivett and Stevens vaccinated mice with a live attenuated strain of *Salmonella dublin,* and the mice were protected against challenge by *Salmonella dublin*, *Salmonella* Typhimurium, *Salmonella choleraesuis,* and *Salmonella anatum*. When day-old chicks were orally vaccinated and subsequently challenged with *Salmonella* Typhimurium, the growth of the challenge organism was considerably reduced or eliminated from the livers of the vaccinated chicks, whereas most of the nonvaccinated were heavily infected (Knivett and Stevens, 1971).

In Europe, a panel (European Food Safety Authority [EFSA], 2004) concluded that vaccines can decrease public health risk caused by *Salmonella* in poultry products by reducing the colonization of reproductive tissues as well as reducing fecal shedding. There is experimental and some limited field evidence that a reduced level of fecal excretion and systemic invasion of *Salmonella* organisms in vaccinated birds will result in a reduced contamination of table eggs and the environment. The authors expressed concern by stating, "Since vaccination cannot guarantee freedom of *Salmonella*, and the consequences of spreading from the top of the pyramid of poultry production would be severe, it is unlikely to be considered in great grand parents of layers and broilers." The authors also mentioned that vaccination does not eliminate the shedding of *Salmonella* (EFSA, 2004).

It is generally accepted that live *Salmonella* vaccines are more effective against infection than are inactivated vaccine preparations (Lillehoj et al., 2007). There are many variables that can affect the efficacy of vaccines, including challenge strain, route of administration, infection dose, parameters used to evaluate the course of infection, age of birds, and species of birds. As reported by the EFSA (2004), researchers have demonstrated that vaccination of chickens results in a considerable quantitative reduced level and duration of intestinal colonization and a diminished systemic invasion by *Salmonella* challenge organisms. The EFSA (2004) reported that, aside from the induction of a strong adaptive immune response, oral administration of live *Salmonella* bacteria to young birds resulted in extensive intestinal colonization, which conferred additional protective effects that are potentially of great value.

Hassan and Curtiss (1997) used an avirulent live *Salmonella* Typhimurium strain to determine its long-term protection efficacy. A comparison of *Salmonella* isolated from vaccinated and nonvaccinated layers after challenge with *Salmonella* Typhimurium or SE showed that the vaccine induced excellent protection against intestinal, visceral, reproductive tract, and egg colonization, invasion, and contamination by *Salmonella* (Hassan and Curtiss, 1997). The authors stated that this was the first publication confirming that vaccination with live avirulent *Salmonella* can induce long-term protection against *Salmonella* infection in layers.

Other scientists (Liu et al., 2001) used formalin-inactivated, microsphere-encapsulated *Salmonella* Enteritidis PT4 to vaccinate specific-pathogen-free chickens in a single dose or vaccinated intramuscularly. When challenged intramuscularly, 10% of the orally vaccinated and 10% of the intramuscularly vaccinated birds showed clinical signs and death, whereas all of the nonvaccinated control birds were sick, and 92% of them were killed. The authors reported that when challenged orally, 26.1% of cloacal swabs and 24.0% of organs from orally vaccinated birds were positive for SE, compared to 27.9% of feces and 18.7% of organs from the intramuscularly vaccinated birds. However, these figures were significantly lower than those for nonvaccinated birds, from which 59.3% of feces and 44.0% of organs tested SE positive. Liu et al. (2001) stated that this was the first evidence that chickens vaccinated with killed SE-loaded poly DL-lactide co-glycolic acid (PLGA) microspheres, intramuscularly and orally in a single dose, developed systematic and local immune responses, thereby conferring protective immunity.

Holt et al. (2003) conducted trials to determine if immunizing specific-pathogen-free, *Salmonella*-culture-negative hens via aerosol exposure to MeganVac1, a commercially available attenuated-*Salmonella* Typhimurium vaccine, would reduce transmission of SE from infected hens to uninfected but contact-exposed hens during a molt. Vaccination reduced the horizontal spread of SE in vaccinated hens compared with their nonvaccinated counterparts, with vaccinated hens shedding significantly less SE. Recovery of SE from ovaries was significantly reduced in the vaccinated hens and from livers/spleens, ovaries, and ceca. The authors observed that immunization of hens with a live *Salmonella* Typhimurium vaccine could help reduce SE problems during a molt situation.

In 2004, Roland et al. developed a live, attenuated vaccine strain to protect chickens against colonization by *Salmonella*. White Leghorn chickens were vaccinated at 1 day of age and after 2 weeks with the test vaccine strains. A nonvaccinated group served as a control. After 4 weeks of growth, all birds were challenged with wild-type *Salmonella hadar* and then necropsied 6 days later. Seventy percent of the nonvaccinates, 60% of the Strain 1 vaccinates, and 15% of the Strain 2 vaccinates were positive for *Salmonella hadar* in tissues (Roland et al., 2004). The number of cecal *Salmonella hadar* isolated from the control group was 1.0×10^6 CFU/g, and from the vaccinated group, this value was 32 CFU/g, indicating a 4 to 5 $\log_{10}$ reduction in colonization by the challenge strain (Roland et al., 2004).

9.7 Effect of Litter

Because chickens are reared on litter on which they defecate and feed, there is a tremendous opportunity for the litter to be implicated as a significant source of *Salmonella* contamination.

9.7.1 Moisture Content of Litter

Many variables can affect the condition of the litter in poultry houses. Leaky waterers, high environmental humidity, poor ventilation, and use of misting systems can all significantly increase the moisture content (MC) of litter. The effect of high moisture in litter can be dramatic regarding the growth and spread of *Salmonella* within the house. Carr et al. (1995) collected litter samples from 24 flocks of broilers, and 4 flocks of broiler breeders were evaluated for *Salmonella* contamination, available moisture (*Aw*), and total MC. On dry litter surfaces, high *Aw* values (0.90–0.95) were associated with *Salmonella*-positive flocks, whereas low *Aw* values (0.79–0.84) were associated with *Salmonella*-negative flocks (Carr, 1995). The authors found that limiting *Aw* in the litter base of broiler houses may create a less-favorable environment for the multiplication of *Salmonella* and thus a more hygienic environment for broiler production.

In 2001, Eriksson de Rezende et al. pointed to the importance of controlling moisture in the grow-out house. The authors found that higher *Salmonella* and *Escherichia coli* counts were detected in litter samples with *Aw* greater than 0.90 and percentage MC greater than 35%. At reduced *Aw* and MC levels, the numbers of viable *Salmonella* cells were low, indicating the importance of preventing excessively damp areas (e.g., "cake") in broiler litter (Eriksson de Rezende et al., 2001).

9.7.2 Effect of pH and Water Activity of Litter

Payne et al. (2007) studied the combined effects of pH and water activity (*Aw*) at a constant temperature on *Salmonella* populations in used turkey litter to predict microbial response over time. Litter was pH adjusted, inoculated with a *Salmonella* cocktail with an initial concentration of 10^7 CFU/g, and placed into individual sealed plastic containers with saturated salt solutions for controlling *Aw*. As litter

Aw and pH levels were reduced, populations declined, with the most drastic reductions (approximately 5 $\log_{10}$ in 9 hours) occurring in low-pH (4) and low-*Aw* (0.84) environments. The authors reported that their findings suggested that the best management practices and litter treatments that lower litter *Aw* to less than or equal to 0.84 and pH to less than or equal to 4 are effective in reducing *Salmonella* populations (Payne et al., 2007).

9.7.3 Effect of Litter Amendments

Evaluations were conducted to determine the efficacy of two litter amendments in reducing or eliminating *Salmonella* in pine litter shavings (Payne et al., 2002). Litter was inoculated with 100 mL of 10^4 CFU/mL nalidixic-acid-resistant *Salmonella* Typhimurium (NAL-SAL) and a sodium bisulfate amendment or a granulated sulfuric acid litter treatment. The recovery rate for *Salmonella* on control litter was 4.4 $\log_{10}$/g; however, no *Salmonella* were recovered from treated litter samples. Litter pH for the control litter was 6.47. Litter treatments decreased the pH from 1.53 to 1.95 depending on the amount applied (Payne et al., 2002).

In 2006, Line and Bailey studied two commercially available litter treatments, aluminum sulfate and sodium bisulfate, to determine their effect on *Campylobacter* and *Salmonella* levels associated with commercial broilers during a 6-week grow-out period. The authors found that, at the application rates investigated, both acidifying litter treatments caused a slight delay in the onset of *Campylobacter* colonization in broiler chicks. *Salmonella* levels remained unaffected using either treatment (Line and Bailey, 2006).

Rothrock et al. (2008) researched the microbiological effect of using alum, which is a common poultry litter amendment that decreases water-soluble phosphorus or reduces ammonia volatilization in poultry litter. Adding alum to poultry litter resulted in significant reductions in both *Campylobacter jejuni* and *Escherichia coli* concentrations by the end of the first month of the experiment. The incidence of *Aspergillus* spp. increased from 0% to 50% of the samples taken over the course of the experiment. The authors concluded that this suggested that the addition of alum to poultry litter potentially shifts the microbial populations from bacterially dominated to those dominated by fungi (Rothrock et al., 2008).

9.8 Effect of Air Movement and Tunnel Ventilation on *Salmonella* Spread

Most of the poultry industry uses tunnel ventilation to cool and ventilate chickens and turkeys (Figure 9.3). This system employs the use of inlets at one end of the poultry house and exhaust fans at the far end of the poultry house. One concern when ventilating a poultry house using this method is that *Salmonella*

Figure 9.3   A tunnel-ventilated broiler chicken house.

may be spread on fomites as air moves rapidly (8.5 to 14.2 m³/minute) through the house. Moreover, opportunities exist for *Salmonella* to be transmitted between laying hens that are held in large buildings in cages. Gast et al. (2004) found that, in rooms containing SE-infected laying hens, air samples were positive for SE for 3 weeks postinoculation. The authors detected *Salmonella* in 66.7% of air samples under these conditions.

Researchers set out to determine whether negative air ionization (also called electrostatic space charging) could affect the airborne transmission of SE (Gast et al., 1999). Groups of chicks were placed in the upstream ends of the mock-scale poultry house cabinets and orally inoculated with SE at 1 week of age. On the following day, 1-day-old chicks were placed in the downstream ends of the cabinets. When chicks were sampled at 3 and 8 days postinoculation, SE was found on the surface of 89.6% of the downstream chicks from cabinets without negative air ionizers, but on only 39.6% of the downstream chicks in the presence of the ionizers (Figure 9.4; Gast et al., 1999). Similarly, SE was recovered from the ceca of 53.1% of sampled downstream chicks in cabinets without ionizers, but from only 1.0% of the ceca of chicks in cabinets in which ionizers were installed (Figure 9.5; Gast et al., 1999). The presence of the ionizers was also associated with reduced levels of circulating airborne dust particles. The authors concluded that reducing airborne dust levels may thus offer an opportunity to limit the spread of SE infections throughout poultry flocks (Gast et al., 1999).

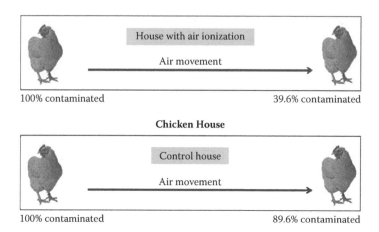

Figure 9.4 Percentage of chicks with *Salmonella* on their feathers in a mock-scale poultry house with or without air ionization. (From From Gast R K, Mitchell B W, and Holt P S, 1999, Application of negative air ionization for reducing experimental airborne transmission of *Salmonella enteritidis* to chicks, *Poultry Science*, 78, 57–61.)

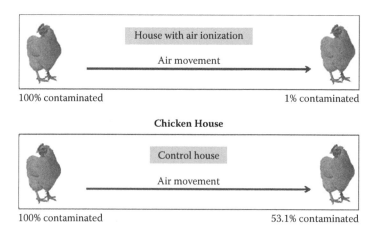

Figure 9.5 Percentage of chicks with *Salmonella* on their ceca in a mock-scale poultry house with or without air ionization. (From From Gast R K, Mitchell B W, and Holt P S, 1999, Application of negative air ionization for reducing experimental airborne transmission of *Salmonella enteritidis* to chicks, *Poultry Science*, 78, 57–61.)

References

Andreatti-Filho R L, Higgins J P, Higgins S E, Gaona G, Wolfenden A D, Tellez G, and Hargis B M (2007), Ability of bacteriophages isolated from different sources to reduce *Salmonella enterica* serovar Enteritidis in vitro and in vivo, *Poultry Science*, 86, 1904–1909.

Bailey J S, Blankenship L C, and Cox N A (1991), Effect of fructooliogsaccharide on *Salmonella* colonization of the chicken intestine, *Poultry Science*, 70, 2433–2438.

Blankenship L C, Bailey J S, Cox N A, Stern N J, Brewer R, and Williams O (1993), Two-step mucosal competitive exclusion flora treatment to diminish salmonellae in commercial broiler chickens, *Poultry Science*, 72, 1667–1672.

Borie C, Albala I, Sanchez P, Sanchez M L, Ramirez S, Navarro C, Morales M A, Retamales J, and Robeson J (2008), Bacteriophage treatment reduces *Salmonella* colonization of infected chickens. *Avian Dis.*, 52(1):64–67.

Carr L E, Mallinson E T, Tate C R, Miller R G, Russek-Cohen E, Stewart L E, Opara O O, and Joseph S W (1995), Prevalence of *Salmonella* in broiler flocks: Effect of litter water activity, house construction, and watering devices, *Avian Diseases*, 39, 39–44.

Collman J P (2001), *Naturally Dangerous: Surprising Facts about Food, Health, and the Environment*, University Science Books, Sausalito, CA, 92.

Corrier D E, Hargis B M, Hinton A, Jr, and Deloach J R (1993a), Protective effect of used poultry litter and lactose in the feed ration on *Salmonella* Enteritidis colonization of leghorn chicks and hens, *Avian Diseases*, 37, 47–52.

Corrier D E, Nisbet D J, Hollister A G, Scanlan C M, Hargis B M, and Deloach J R (1993b), Development of defined cultures of indigenous cecal bacteria to control salmonellosis in broiler chicks, *Poultry Science*, 72, 1164–1168.

Cox N A, Bailey J S, Blankenship L C, and Gildersleeve R P (1992), In ovo administration of a competitive exclusion culture treatment to broiler embryos, *Poultry Science*, 71, 1781–1784.

Eriksson De Rezende C L, Mallinson E T, Tablante N L, Morales R, Park A, Carr L E, and Joseph S W (2001), Effect of dry litter and airflow in reducing *Salmonella* and *Escherichia coli* populations in the broiler production environment, *Journal of Applied Poultry Research,* 10, 245–251.

European Food Safety Authority (2004), Opinion of the Scientific Panel on Biological Hazards on a request from the commission related to the use of vaccines for the control of *Salmonella* in poultry, *Journal of the European Food Safety Authority*, 114, 1–74.

Fiorentin L, Vieira N D, and Barioni W, Jr (2005), Oral treatment with bacteriophages reduces the concentration of *Salmonella* Enteritidis PT4 in caecal contents of broilers, *Avian Pathology,* 34, 258–263.

Gast R K, Mitchell B W, and Holt P S (1999), Application of negative air ionization for reducing experimental airborne transmission of *Salmonella* Enteritidis to chicks, *Poultry Science*, 78, 57–61.

Gast R K, Mitchell B W, and Holt P S (2004), Evaluation of culture media for detecting airborne *Salmonella* Enteritidis collected with an electrostatic sampling device from the environment of experimentally infected laying hens, *Poultry Science,* 83, 1106–1111.

Hassan J O, and Curtiss R, III (1997), Efficacy of a live avirulent *Salmonella* Typhimurium vaccine in preventing colonization and invasion of laying hens by *Salmonella* Typhimurium and *Salmonella* Enteritidis, *Avian Diseases,* 41, 783–791.

Hollister A G, Corrier D E, Nisbet D J, and Deloach J R (1994), Effect of cecal cultures encapsulated in alginate beads or lyophilized in skim milk and dietary lactose on *Salmonella* colonization in broiler chicks, *Poultry Science,* 73, 99–105.

Holt P S, Gast R K, and Kelly-Aehle S (2003), Use of a live attenuated *Salmonella* Typhimurium vaccine to protect hens against *Salmonella* Enteritidis infection while undergoing molt, *Avian Diseases,* 47, 656–661.

Impey C S (1982), Competitive exclusion of salmonellas from the chick caecum using a defined mixture of bacterial isolates from the caecal microflora of an adult bird, *Journal of Hygiene,* 89, 479–490.

Knivett V A, and Stevens W K (1971), The evaluation of a live *Salmonella* vaccine in mice and chickens, *Journal of Hygiene,* 69, 233–245.

Lillehoj H S, Kim C H, Keeler C L, Jr, and Zhang S (2007), Immunogenomic approaches to study host immunity to enteric pathogens, *Poultry Science,* 86, 1491–1500.

Line J E, and Bailey J S (2006), Effect of on-farm litter acidification treatments on *Campylobacter* and *Salmonella* populations in commercial broiler houses in northeast Georgia, *Poultry Science,* 85, 1529–1534.

Line J E, Bailey J S, Cox N A, and Stern N J (1997), Yeast treatment to reduce *Salmonella* and *Campylobacter* populations associated with broiler chickens subjected to transport stress, *Poultry Science,* 76, 1227–1231.

Liu W, Yang Y, Chung N, and Kwang J (2001), Induction of humoral immune response and protective immunity in chickens against *Salmonella* Enteritidis after a single dose of killed bacterium-loaded microspheres, *Avian Diseases,* 45, 797–806.

McGrath S, and Van Sinderen D, eds (2007), *Bacteriophage: Genetics and Molecular Biology,* Caister Academic Press, Hethersett, UK.

Nurmi E, Johansson T, and Nuotio L (1992), Long-term experience with competitive exclusion and salmonellas in Finland, *International Journal of Food Microbiology,* 15, 281–285.

Payne J B, Kroger E C, and Watkins S E (2002), Evaluation of litter treatments on *Salmonella* recovery from poultry litter, *Journal of Applied Poultry Research,* 11, 239–243.

Payne J B, Osborne J A, Jenkins P K, and Sheldon B W (2007), Modeling the growth and death kinetics of *Salmonella* in poultry litter as a function of pH and water activity, *Poultry Science,* 86, 191–201.

Rantala M, and Nurmi E (1973), Prevention of growth of *Salmonella infantis* in chicks by the flora of the alimentary tract of chickens, *British Poultry Science,* 14, 627–630.

Roland K, Tinge S, Warner E, and Sizemore D (2004), Comparison of different attenuation strategies in development of a *Salmonella hadar* vaccine, *Avian Diseases,* 48, 445–452.

Rothrock M J, Jr, Sistani K, Warren J G, and Cook K L (2008), The effect of alum addition on microbial communities in poultry litter, *Poultry Science,* 87, 1493–1503.

Thitaram S N, Chung C H, Day D F, Hinton A, Jr, Bailey J S, and Siragusa G R (2005), Isomaltooligosaccharide increases cecal *Bifidobacterium* population in young broiler chickens, *Poultry Science,* 84, 998–1003.

Toro H, Price S B, McKee S, Hoerr F J, Krehling J, Perdue M, and Bauermeister L (2005), Use of bacteriophages in combination with competitive exclusion to reduce *Salmonella* from infected chickens, *Avian Diseases,* 49, 118–124.

Watarai S, and Tana (2005), Eliminating the carriage of *Salmonella enterica* serovar Enteritidis in domestic fowls by feeding activated charcoal from bark containing wood vinegar liquid (Nekka-Rich), *Poultry Science*, 84, 515–521.

Wierup M, Wahlstrom H, and Engstrom B (1992), Experience of a 10-year use of competitive exclusion treatment as part of the *Salmonella* control programme in Sweden, *International Journal of Food Microbiology*, 15, 287–291.

Chapter 10

Effect of the Health of Chickens on *Salmonella* Prevalence

10.1 Introduction

The influence of coccidiosis on colonization of *Salmonella* Typhimurium in broiler chickens under floor pen conditions was studied by Arakawa et al. (1992). Chickens were separated into two groups, unmedicated and medicated with nicarbazin, and exposed to three species of *Eimeria* (*Eimeria tenella*, *Eimeria maxima*, and *Eimeria acervulina*) at 2, 3, and 4 weeks of age and given *Salmonella* Typhimurium in the feed 2 days later. *Salmonella* Typhimurium was isolated most often (100%) from ceca of chickens exposed at 3 weeks of age. The most significant finding was that birds in the unmedicated group were positive for *Salmonella* Typhimurium at a higher rate than those in the medicated group. *Salmonella* Typhimurium was detected in livers only in a few unmedicated birds. Thus, disease conditions can contribute to colonization of the bird by *Salmonella*.

In a more indirect way, diseases that affect the integrity of the intestinal tract can also have an impact on fecal contamination, which may affect contamination of the carcass by *Salmonella* or *Campylobacter* during processing. The National Advisory Committee on Microbiological Criteria for Foods (NACMCF, 1997) reported that because processing of raw broilers does not involve a lethal heat process, such as pasteurization, delivering live chickens to the processing plant with as few pathogens as possible is necessary to control contamination of carcasses with *Salmonella* and *Campylobacter*. Other scientists have supported this conclusion by stating that reducing *Campylobacter jejuni* colonization in live chickens should reduce the prevalence of *C. jejuni* infections in humans, presumably because of less exposure to the organism (Morishita et al., 1997). Controlling factors that contribute to colonization of the live bird during grow out should have a significant impact on contamination of finished carcasses after processing.

10.2 Effect of Intestinal Damage

At the processing plant, fecal contamination of carcasses can become a concern if the digestive tracts of chickens are cut or torn during venting, opening, or evisceration because, if cut or torn, fecal material may be released onto the surface of the carcass (NACMCF, 1997). Intestinal damage is generally associated with improperly adjusted or worn-out evisceration equipment, variance among individual birds, or birds with low body weight due to disease. These factors must be controlled because modern poultry-processing plants are highly automated operations, and the equipment is set to receive carcasses of a specific size. Preventing contamination of carcasses from spillage of digestive tract contents or smearing of fecal material on edible meat surfaces is perhaps the single most important factor in sanitary poultry slaughter (Bilgili, 2001). If intestines are cut or torn during evisceration, feces can spread to equipment, workers, and inspectors and can be a major source of cross contamination with pathogenic bacteria (NACMCF, 1997).

10.3 Effect of Air Sacculitis

Pilot studies conducted in 1997 by two vertically integrated broiler companies revealed a direct relationship between air sacculitis infections in chickens and the presence of high numbers of *Escherichia coli* (*E. coli*) and *Salmonella*. In the first study, carcasses that were removed from the line by the Food Safety Inspection Service (FSIS) of the U.S. Department of Agriculture (USDA) inspectors for active air sacculitis infections and carcasses that were not visibly infected were evaluated for *E. coli* counts over a 1-week period (unpublished data, a commercial poultry facility, 2001). For carcasses with no visible signs of air sacculitis, 58% had prechill *E. coli* numbers in the acceptable range (0 to <100 CFU [colony-forming units]/ mL) according to the hazard analysis and critical control point (HACCP) regula­tion (USDA-FSIS, 1996), 37% were in the questionable range (100 to 1,000 CFU/ mL), and only 5% were in the unacceptable range (>1,000 CFU/mL; Figure 10.1). However, for carcasses removed from the line for air sacculitis infections, 4%, 46%,

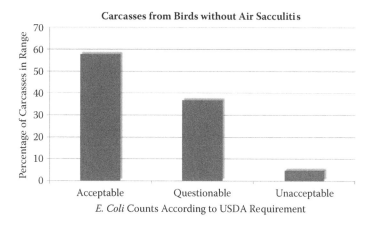

Figure 10.1   Percentage of carcasses in a specific USDA-FSIS range for *E. coli* for birds that did not have active air sacculitis infections.

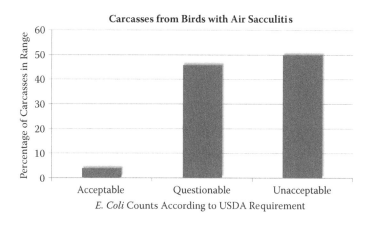

Figure 10.2   Percentage of carcasses in a specific USDA-FSIS range for *E. coli* for birds with active air sacculitis infections.

and 50% of the prechill carcasses were in the acceptable, questionable, and unacceptable ranges, respectively. Therefore, a total of 96% of air sacculitis-infected carcasses had questionable or unacceptable *E. coli* counts (Figure 10.2).

In a second study conducted by another integrator, prechill *E. coli* counts for carcasses with air sacculitis were higher (at 3.93 $\log_{10}$ CFU/mL) than air sacculitis-negative carcasses at 2.63 $\log_{10}$ CFU/mL. Moreover, this company found that *Salmonella* prevalence for carcasses with air sacculitis was significantly higher at 70% than for carcasses without air sacculitis at 40% (unpublished data, a

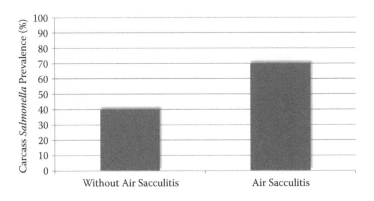

Figure 10.3 Carcass *Salmonella* prevalence for carcasses from chickens without or with active air sacculitis infections.

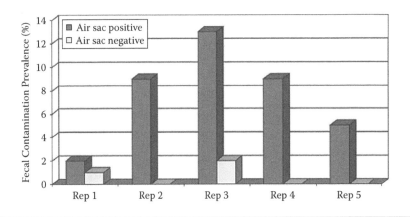

Figure 10.4 Prevalence of fecal contamination on carcasses of birds that were air sacculitis positive or negative. (From Russell S M, 2003, The effect of air sacculitis on bird weights, uniformity, fecal contamination, processing errors, and populations of *Campylobacter* spp and *Escherichia coli*, *Poultry Science*, 82, 1326–1331.)

commercial processing facility, 2001). These studies demonstrated a link between the presence of air sacculitis in the flock and increases in indicator and pathogenic bacterial populations (Figure 10.3).

Russell (2003) conducted a study to find out if air sacculitis infections in broiler chickens have an effect on the percentage of carcasses with fecal contamination and on populations of *Campylobacter* and *E. coli*. Air sacculitis-positive flocks were compared with air sacculitis-negative flocks for the factors listed. The author reported that air sacculitis infections significantly increased fecal contamination in four of five repetitions (Figure 10.4). Thus, it is reasonable to conclude that air sacculitis in

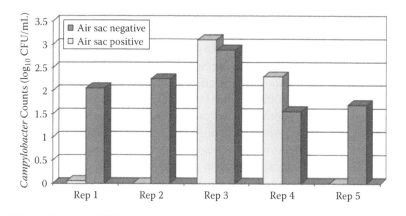

Figure 10.5 The effect of air sacculitis on *Campylobacter* counts on broiler chicken carcasses. (From Russell S M, 2003, The effect of air sacculitis on bird weights, uniformity, fecal contamination, processing errors, and populations of *Campylobacter* spp and *Escherichia coli*, *Poultry Science*, 82, 1326–1331.)

flocks of birds may contribute to increases in the risk for human food-borne infection with *Salmonella* and *Campylobacter*.

This study also found that *Campylobacter* counts were significantly higher on carcasses from air sacculitis flocks in three of five repetitions (Figure 10.5).

In all cases where air sacculitis infections were present, *Campylobacter* was present at levels greater than 1.5 $\log_{10}$ CFU/mL. In three of the repetitions, when air sacculitis was absent, *Camplyobacter* counts were absent or extremely low (<0.05 $\log_{10}$ CFU/mL). Therefore, there was a relationship between the presence of air sacculitis and *Campylobacter*-positive carcasses (Russell, 2003).

The data collected in this study suggested that flocks of chickens with air sacculitis infections are more likely to be contaminated with fecal material during processing and have higher *Campylobacter* counts.

This study led to an analysis to determine if there is a relationship between air sacculitis infections, fecal contamination, and indicator and pathogenic bacterial counts in commercial poultry-processing plants. A commercial poultry processor allowed its records to be reviewed regarding specific processing parameters for 32.3 million chickens processed over a 2-year period. The data were analyzed by the Department of Statistics at the University of Georgia (unpublished data, 2002). The analyses showed that as the percentage of carcasses removed from the line for air sacculitis disease by the USDA inspectors increased, the percentage of carcasses with fecal contamination increased as well. Increasing levels of infectious process also resulted in a significant increase in fecal contamination. The data revealed that when a high number of carcasses are condemned, increased fecal contamination occurred. A significant finding was that, as the number of carcasses removed from the line for active air sacculitis infection increased, the prevalence of *Salmonella* on processed carcasses increased as well. The statistician concluded that "with samples of the size used in this investigation, these differences are quite significant; there is very convincing evidence that air sacculitis increase is associated with increasing probability of *Salmonella* contamination." This conclusion strongly supported the research of Russell (2003), which found that the presence of air sacculitis increases *Campylobacter* populations on carcasses, and the research obtained from the previous pilot studies. As air sacculitis percentage increased, carcasses with an infectious process also increased in a progressive manner. Thus, disease conditions such as coccidiosis, air sacculitis, and an inflammatory or infectious process can significantly increase colonization of chickens with pathogens such as *Salmonella* or *Campylobacter*.

10.4 Effect of Transport Coops

Another source of potential contamination is the transport coops used to deliver birds to the processing plant. In 1998, Bolder reported that excretion of pathogenic bacteria while birds are being transported to the processing plant might cause cross contamination among birds during the trip. Altekruse (1998) found that transportation of broilers to the processing plant caused bacterial counts to increase up to 1,000-fold. Rigby et al. (1982) isolated *Salmonella* from 99% of the empty transport coops that were evaluated (98/99 coops) before the chickens were loaded. These same authors attributed *Salmonella* contamination of a previously negative flock to the transport coops (Rigby et al., 1982).

10.5 Effect of Floor Type

In 2000, Buhr et al. conducted four trials to determine whether conventional solid or elevated wire mesh flooring, during transport and holding of broilers prior to slaughter, influenced the number of bacteria recovered from feathered and defeathered carcasses. In this study, broilers were transported for 1 hour and then held for 13 hours under a covered shed before processing. The authors found that although broilers transported and held on solid flooring had noticeably dirtier breast feathers and higher coliform and *E. coli* counts prior to scalding and defeathering, bacterial recovery from external carcass rinses did not differ between the solid and wire flooring treatments after defeathering (Buhr et al., 2000).

The prevalence and types of *Salmonella* in broiler chickens during transportation and during slaughter and dressing were studied by Corry et al. (2002). Transportation coops were contaminated with *Salmonella* after washing and disinfection. Corry et al. (2002) found that the reasons for transport coop contamination were (1) inadequate cleaning, resulting in residual fecal soiling; (2) too low disinfectant concentration and temperature of disinfectant; and (3) contaminated recycled flume water used to soak the coops.

10.6 Effect of Transportation

In another study, Ramesh et al. (2003) studied coops used in transporting live broilers between the grow-out houses and processing plants and found that they are a primary source of contamination for processed poultry products. Because disinfection of transport coops has been difficult to accomplish, it is probable that the choice of appropriate disinfectant and its application are partially or wholly responsible for the failure to adequately eliminate pathogens from the coops (Ramesh et al., 2003). The authors observed that two disinfectants used under simulated caked-on feces conditions suggested that application under the prescribed regimen could result in effective elimination of *Salmonella* from transport coops within a limited period. In another study in 2003, Ramesh et al. used heat and chlorine to disinfect poultry transport coops successfully.

References

Altekruse S F (1998), *Campylobacter jejuni* in foods, *Veterinary Medicine Today*, 213, 1734–1735.

Arakawa A, Fukata T, Baba E, McDougald L R, Bailey J S, and Blankenship L C (1992), Influence of coccidiosis on *Salmonella* colonization in broiler chickens under floor-pen conditions, *Poultry Science*, 71, 59–63.

Bilgili S F (2001), Poultry meat inspection and grading, in *Poultry Meat Processing*, ed A Sams, CRC Press, Boca Raton, FL, 47–71.

Buhr R J, Cason J A, Dickens J A, Hinton Jr A, and Ingram K D (2000), Influence of flooring type during transport and holding on bacteria recovery from broiler carcass rinses before and after defeathering, *Poultry Science*, 79, 436–441.

Corry J E L, Allen V M, Hudson W R, Breslin M F, and Davies R H (2002), Sources of *Salmonella* on broiler carcasses during transportation and processing: modes of contamination and methods of control, *Journal of Applied Microbiology*, 92, 424–432.

Morishita T Y, Aye P P, Harr B S, Cobb C W, and Clifford J R (1997), Evaluation of an avian-specific probiotic to reduce the colonization and shedding of *Campylobacter jejuni* in broilers, *Avian Diseases*, 41, 850–855.

National Advisory Committee on Microbiological Criteria for Foods (1997), Generic HACCP application in broiler slaughter and processing, *Journal of Food Protection*, 60, 579–604.

Ramesh N, Joseph S W, Carr L E, Douglass L W, and Wheaton F W (2003), Serial disinfection with heat and chlorine to reduce microorganism populations on poultry transport containers, *Journal of Food Protection*, 66, 793–797.

Rigby C E, Petit J R, Bently A H, Spencer J L, Salomons M O, and Lior H (1982), The relationship of salmonellae from infected broiler flocks, transport crates or processing plants to contamination of eviscerated carcasses, *Canadian Journal of Comparative Medicine*, 46, 272–278.

Russell S M (2003), The effect of air sacculitis on bird weights, uniformity, fecal contamination, processing errors, and populations of *Campylobacter* spp and *Escherichia coli*, *Poultry Science*, 82, 1326–1331.

United States Department of Agriculture (1996), Pathogen reduction, hazard analysis and critical point (HACCP) systems; Final rule. *Code of Federal Regulations 9*, part 304, 38867.

Chapter 11

Sources of *Salmonella* in the Plant

11.1 Introduction

Salmonella enter the processing plant on the birds and may be spread to equipment during specific processing steps. This chapter details the locations where *Salmonella* may be found on the birds and certain processes that can enhance cross contamination during processing.

11.2 Location of *Salmonella* on Poultry Carcasses

In 2007, Cason et al. evaluated internal and external portions of broiler chickens to determine where and how many *Salmonella* could be recovered from these locations (Table 11.1 and Figure 11.1). These data demonstrated that both *Salmonella* and *Campylobacter* were widely distributed on the chicken and supported the idea of crop disinfection during feed withdrawal.

Table 11.1   *Salmonella* **Prevalence in External and Internal Samples
from Broiler Chickens Just before Processing in 40-mL Qualitative Tests**

			Salmonella log_{10} (MPN)	
Sample	*Qualitative* Salmonella	Salmonella *in Carcasses (%)*	*Per Sample*	*Per Gram*
Feathers	42	60	3.8	2.0
Picked carcass	27	39	3.6	1.5
Head-feet	32	46	3.1	1.9
Ceca	17	24	3.6	2.6
Colon	16	23	3.1	2.7
Crop	14	20	4.0	3.1

Source: From Cason J A, Ingram K D, Smith D P, Cox N A, Hinton A, Jr, Northcutt J A, and Buhr R J, 2007, Partitioning of external and internal bacteria carried by broiler chickens before processing, *Journal of Food Protection*, 70, 2056–2062.

11.3 Influence of the Crop in *Salmonella* Contamination of Carcasses

As mentioned in Chapter 8.17.1, Byrd et al. (2002) reported that research has identified the crop as a source of *Salmonella* and *Campylobacter* contamination for broiler carcasses, and broiler crops are 86 times more likely to rupture than ceca during commercial processing. Using a fluorescent marker at commercial processing plants, the authors evaluated leakage of crop and upper gastrointestinal contents from broilers. Broilers were orally gavaged with a fluorescent marker paste (cornmeal-fluorescein dye-agar) within 30 minutes of live hang (Byrd et al., 2002). Carcasses were collected at several points during processing and were examined for upper gastrointestinal leakage using long-wavelength black light. This survey indicated that 67% of the total broiler carcasses were positive for the marker at the rehang station following head and shank removal. Crops were mechanically removed from 61% of the carcasses prior to the cropper, and visual online examination indicated leakage of crop contents following crop removal by the Pac-Man. Examination of the carcasses prior to the cropper detected the marker in the following regions: neck (50.5% positive), thoracic inlet (69.7% positive), thoracic cavity (35.4% positive), and abdominal cavity (34.3% positive). Immediately prior to chill immersion, 53.2% of the carcasses contained some degree of visually identifiable marker contamination, in areas as follows: neck (41.5% positive), thoracic inlet (45.2% positive), thoracic cavity (26.2% positive), and abdominal cavity (30.2% positive) (Byrd et al., 2002).

Feathers
Coliforms: 5.4
E. coli: 5.1
Salmonella: 2.0
Campylobacter: 4.7

Picked Carcass
Coliforms: 4.1
E. coli: 3.8
Salmonella: 1.5
Campylobacter: 3.6

Ceca
Coliforms: 6.7
E. coli: 6.2
Salmonella: 2.6
Campylobacter: 7.0

Colon
Coliforms: 6.0
E. coli: 5.5
Salmonella: 2.7
Campylobacter: 6.5

Crop
Coliforms: 3.3
E. coli: 2.8
Salmonella: 3.1
Campylobacter: 4.5

Head and Feet
Coliforms: 4.0
E. coli: 3.6
Salmonella: 1.9
Campylobacter: 3.9

Figure 11.1   Distribution of bacterial types on or in a broiler chicken's feathers, carcass after picking, ceca, colon, crop, and head and feet. (Figure used with permission from Bill Marler.)

The authors demonstrated the importance of the cropping machine in spreading *Salmonella* to the internal and external portions of the carcass.

11.4 Effect of Defeathering on Cross Contamination

In 2007, Nde et al. found that *Salmonella* on the feathers of live birds may be transferred to carcass skin during defeathering. In this study, the possibility of transfer of *Salmonella* from the feathers of live turkeys to carcass tissue during the defeathering process at a commercial turkey-processing plant was investigated (Nde et al., 2007). The contribution of scald water and picker fingers to cross contamination was also examined. On four different occasions, swab samples were collected from 174 randomly selected tagged birds before and after defeathering. Two swab samples from picker fingers and a sample of scald water were collected during each visit. Nde et al. (2007) found that *Salmonella* prevalence was similar before and after defeathering during Visits 2 and 3 and significantly increased after defeathering during Visits 1 and 4. Over the four visits, all *Salmonella* subtypes obtained after defeathering were also isolated before defeathering. The results of this study suggest that *Salmonella* was transferred from the feathers to carcass skin during each visit. The authors found that, on each visit, the *Salmonella* subtypes isolated from the fingers of the picker machines were similar to subtypes isolated before and after defeathering, indicating that the fingers facilitated carcass cross contamination during defeathering. *Salmonella* isolated from scald water during Visit 4 was related to isolates obtained before and after defeathering, suggesting that scald water is also a vehicle for cross contamination during defeathering. Using molecular subtyping, Nde et al. (2007) demonstrated the relationship between *Salmonella* present on the feathers of live turkeys and carcass skin after defeathering, suggesting that decontamination procedures applied to the external surfaces of live turkeys could reduce *Salmonella* cross contamination during defeathering.

References

Byrd J A, Hargis B M, Corrier D E, Brewer R L, Caldwell D J, Bailey R H, Mcreynolds J L, Herron K L, and Stanker L H (2002), Fluorescent marker for the detection of crop and upper gastrointestinal leakage in poultry processing plants, *Poultry Science*, 81, 70–74.

Cason J A, Ingram K D, Smith D P, Cox N A, Hinton A, Jr, Northcutt J A, and Buhr R J (2007), Partitioning of external and internal bacteria carried by broiler chickens before processing, *Journal of Food Protection*, 70, 2056–2062.

Nde C W, McEvoy J M, Sherwood J S, and Logue C M (2007), Cross-contamination of turkey carcasses by *Salmonella* species during defeathering, *Poultry Science*, 86, 162–167.

Chapter 12

The Role of the Scalder in Spreading *Salmonella*

12.1 Introduction

Poultry scalders are often implicated in the spread of *Salmonella* from carcass to carcass. This is the first location during processing where chickens are exposed to a common bath, allowing *Salmonella* cells from positive carcasses to spread *Salmonella* to negative carcasses.

12.2 How *Salmonella* Are Released during Scalding

Cason et al. (2006) studied the release of bacteria from individual broiler carcasses in warm water using a scalder model. Immediately after shackling and electrocution, feathered and genetically featherless broiler carcasses were immersed individually in 42°C or air-agitated tap water for 150 seconds. Although any visible fecal material expelled as a result of electrocution was removed before sampling, carcass condition was typical for market-age broilers subjected to 12 hours of feed withdrawal (Cason et al., 2006). Duplicate water samples were taken at 10, 30, 70, 110, and 150 seconds, and *Escherichia coli* counts were determined. Samples of initial tap water and contaminated water approximately 2 minutes after removal of carcasses indicated that *E. coli* could not be detected in the original water source, and that mortality of *E. coli* in the warm water was negligible. Mean numbers of *E. coli* released were 6.2 and 5.5 $\log_{10}$ (colony-forming units [CFU]/carcass) at 150 seconds for feathered and featherless carcasses, respectively. For both feathered and featherless carcasses, the rate of release of *E. coli* was highest in the first 10 seconds, and the rate declined steadily during the remaining sampling period. *Salmonella* release into the scalder water should follow a similar pattern. The authors reported that the findings were compatible with published reports about operating multiple-tank scalders, indicating that a high proportion of total bacteria in a multiple-tank scalder is in the first scald tank that the carcasses enter. Higher numbers of *E. coli* released from feathered carcasses are probably due to the much greater surface area of contaminated feathers compared with the skin of featherless carcasses (Cason et al., 2006).

In 2001, Yang et al. inoculated *Salmonella* Typhimurium and *Campylobacter jejuni* into scald water and chiller water and on chicken skins to determine the effects of scalding temperature (50°C, 55°C, and 60°C) and the chlorine level in chilled water (0, 10, 30, and 50 ppm), associated with the ages of scalding water (0 and 10 hours) and chiller water (0 and 8 hours), on bacterial survival or death. After scalding at 50°C and 60°C, the reductions of *C. jejuni* were 1.5 and 6.2 $\log_{10}$ CFU/mL in water and less than 1 and more than 2 $\log_{10}$ CFU/cm^2, respectively, on chicken skins. The reductions of *Salmonella* Typhimurium were less than 0.5 and more than 5.5 $\log_{10}$ CFU/mL in water and less than 0.5 and more than 2 $\log_{10}$ CFU/cm^2, respectively, on skins (Yang et al., 2001). The age of scalding water did not significantly ($p > 0.05$) affect bacterial heat sensitivity. However, the increase in the age of chilled water significantly ($p < 0.05$) reduced the effect of the chlorine on bacteria. In 0-hour chilled water, *C. jejuni* and *Salmonella* Typhimurium were reduced by 3.3 and 0.7 $\log_{10}$ CFU/mL, respectively, after treatment with 10 ppm of chlorine and became undetectable with 30 and 50 ppm of chlorine, respectively (Yang et al., 2001). In 8-hour chilled water, the reduction of *C. jejuni* and *Salmonella* Typhimurium was less than 0.5 $\log_{10}$ CFU/mL with 10 ppm of chlorine and ranged from 4 to 5.5 $\log_{10}$ CFU/mL with 50 ppm of chlorine. Yang et al. (2001) found that chlorination of chilled water did not effectively reduce the bacteria attached on chicken skins.

Fries (2002) stated that from primary production and on through processing, the microflora on poultry while they are in the grow-out house will be transferred into the processing plant and onto exterior parts of the carcasses. Cross contamination with *Salmonella* occurs during scalding, defeathering, head pulling, evisceration, and chilling (Fries, 2002). He reported that the poultry-processing line does not have any killing capability for *Salmonella*. This is true in European plants and facilities that export poultry to the European Union, but this is not the case in the United States.

The prevalence of *Salmonella* on transport coop surfaces, in water, and on broiler chicken (carcasses, parts, and viscera) taken from a poultry-processing plant located in the southern part of Brazil was studied by Reiter et al. (2007). The authors detected *Salmonella* in the locations presented in Figure 12.1. These data

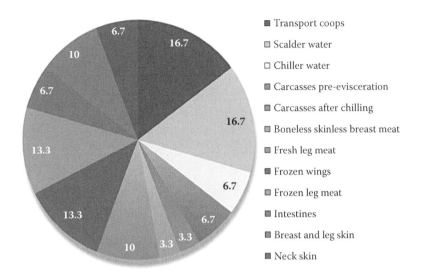

Figure 12.1   Percentage of samples positive for *Salmonella*. (From Reiter M G R, Fiorese M R, Moretto G, Lopez M C, and Jordano R, 2007, Prevalence of *Salmonella* in a poultry slaughterhouse, *Journal of Food Protection*, 70, 1723–1725.)

demonstrated that transport coops, scalder water, chiller water, frozen wings, and frozen leg meat were highly contaminated with *Salmonella*. It is important to note that this study was conducted in Brazil, where no chemicals are used as an intervention because the product is exported to Europe.

References

Cason J A, Buhr R J, and Hinton A, Jr (2006), Release of *Escherichia coli* from feathered and featherless broiler carcasses in warm water, *Poultry Science*, 85, 1807–1810.

Fries R (2002), Reducing *Salmonella* transfer during industrial poultry meat production, *World's Poultry Science Journal*, 58, 527–540.

Reiter M G R, Fiorese M R, Moretto G, Lopez M C, and Jordano R (2007), Prevalence of *Salmonella* in a poultry slaughterhouse, *Journal of Food Protection*, 70, 1723–1725.

Yang H, Li Y, and Johnson M G (2001), Survival and death of *Salmonella* Typhimurium and *Campylobacter jejuni* in processing water and on chicken skin during poultry scalding and chilling, *Journal of Food Protection*, 64, 770–776.

Chapter 13

Controlling *Salmonella* in Poultry Scalders

13.1 Introduction

The immersion scalder (Figure 13.1) is the first place in the poultry-processing plant where fecal material is able to come off the birds and into the surrounding water. As such, the scalder may be a significant source of cross contamination.

13.2 Controlling Fecal Contamination of Scalders

In the southeastern United States in the summertime, companies often experience problems keeping the birds cool in the grow-out house. As a result, some growers will use foggers to wet the birds and keep them cool. These fogging systems often wet the birds, allowing litter to attach to feathers. This may result in a serious problem once the birds arrive at the processing plant because the litter on their feathers and skin is loosened during scalding and comes off the birds, resulting in excessive organic material in the scald water. In addition, in the winter, when houses are closed and ventilation is decreased to lower the energy cost of heating the birds, birds may become caked with fecal material as the litter retains more moisture than when the ventilation system is running. In both cases, excessive fecal material on the birds' feathers (Figures 13.2 and 13.3) is washed off into the scalder water. Scalder water containing high concentrations of fecal material is a problem

145

Figure 13.1   Commercial three-stage immersion poultry scalder.

Figure 13.2   Excessive fecal material clumped onto the breasts of live chickens before scalding.

Figure 13.3 Excessive fecal material clumped onto the breasts of live chickens after scalding.

because it comes in contact with the external surface of the birds, and during picking, bacteria contained in this dirty water may be massaged into the skin and open feather follicles. Also, this organic material may be retained on the surface of the bird through evisceration and end up in the chiller, deactivating chlorine and preventing disinfection.

To reduce this problem, some companies have installed a bird brush and washer (Figures 13.4 and 13.5) prior to scalding. Larger brushes and chlorinated water physically remove the feces from the feathers and skin of the birds. One company using this technique decreased the amount of fecal material going into the scalder by approximately 90%. This decreases the amount of organic material on the surface of carcasses as they go into the chiller.

To achieve the best results possible, a bird brush system that allows for the scrubbing of the front, back, and saddle of each bird is advisable. This is depicted in Figure 13.6.

The next important step in removing organic material from carcasses is the scalder. The scalder is one of the most important areas in the processing plant in which cross contamination with *Salmonella* can occur. Most scalders are not set up to be truly countercurrent. The water should move against the carcasses, going from the exit of the scalder toward the entrance. This opposing water flow is essential to wash the birds and remove contamination from the birds as they travel through the scalder. Countercurrent flow may be accomplished by adding a steel

Figure 13.4   Outside view of a prescald bird brush system.

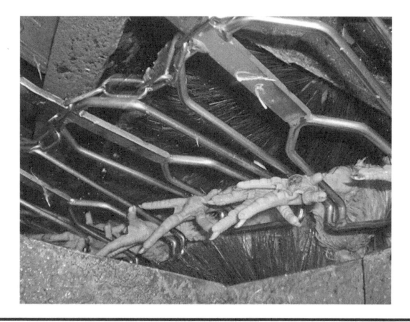

Figure 13.5   Inside view of a prescald bird brush system.

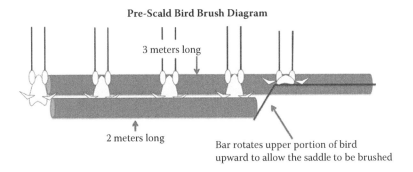

Figure 13.6   **Prescald bird brush diagram.**

barrier between the lines of chickens going in either direction. By separating these chickens, bacteria that are washed off the external surface of the chickens entering the scalder are not transferred to those exiting the scalder. The rate of water flow should be high to dilute the concentration of foreign material and bacteria in the scalder. Usually, greater than 1 L per bird is recommended. There is a common adage that "dilution is the solution to pollution," and it applies in this case. Plants that are not equipped with multistage scalders (scalders with successive, separate tanks) should attempt to make their scalders multistage.

13.3 *Salmonella* Attachment during Scalding

Kim et al. (1993) studied the microtopography of chicken skin exposed to varying scalding temperatures to determine the least-favorable skin surface for salmonellae attachment. Birds were scalded at 52°C, 56°C, and 60°C, and the changes of skin morphology were examined by light and transmission electron microscopy throughout the processing operation. Breast skins obtained immediately after picking were inoculated with *Salmonella* Typhimurium, and the attachment was quantified using scanning electron microscopy and microbiological plating techniques. Chicken skins scalded at 52°C and 56°C retained most of the epidermis, although the latter temperature caused the loss of twice as much stratum corneum layers and produced a smoother surface than the former (Kim et al., 1993). Skins at 60°C began to lose most of epidermal layers during scalding and exposed the dermal surface of the skin after picking, which was sometimes covered with thin fragmental epidermis or basal tissue. The number of salmonellae attached to 60°C processed skins was 1.1 to approximately 1.3 $\log_{10}$ higher than those attached to the skins processed at 52°C and 56°C, as measured by scanning electron microscopy. Microbiological plating, however, showed no significant difference in attachment among three skins processed at different temperatures. The authors thought that this was due to the insensitivity of the plating method to differentiate attachment strengths of salmonellae to the skin. Kim et al. (1993) concluded that removal of the entire epidermis should be avoided in processing to reduce salmonellae attachment to the skin. Thus, soft scalded chicken should be more resistant to attachment by *Salmonella* when compared to hard scalded birds.

13.4 Effect of Scalding on Bacterial Numbers

In 1994, Abu-Ruwaida et al. studied the effect of processing procedures on the overall microbiological quality and safety of chicken carcasses. The authors found that the highest bacterial levels were detected after scalding and defeathering. Bryan and Doyle (1995) stated that scalding, defeathering, evisceration, and giblet operations are major points of spread with *Salmonella*.

Cason et al. (1999) collected scald water and whole carcass rinses on nine different days in a commercial broiler-processing plant operating adjacent lines that processed birds from the same flock simultaneously. A conventional, single-tank, two-pass scalder was installed on one line, and the other line had a three-tank, two-pass, counterflow scalder in which water mixed across the two lines of carcasses within each tank. Water samples from the turnaround point in each tank were analyzed for aerobic bacteria and suspended solids. At the same time that water samples were taken, six carcasses were removed from the processing line immediately after feather removal and rinsed. Estimated numbers of aerobic bacteria were significantly reduced in the third tank of the counterflow scalder compared to the second tank or compared to the single tank of the conventional scalder. Despite the differences in aerobic bacteria between scald tanks, numbers of aerobic bacteria in carcass rinses were not affected by scalder design. Organic and total solids were significantly reduced in the third tank of the counterflow scalder compared to the first and second tanks and in the third tank of the counterflow scalder compared to the conventional scalder. Solids in the third (final) tank of the counterflow scalder were reduced by about 70% compared to the conventional scalder.

In 2000, Cason et al. studied water samples from a commercial broiler-processing plant scalder and tested them for coliforms, *E. coli*, and *Salmonella* to evaluate the numbers of suspended bacteria in a multiple-tank, counterflow scalder. Sixteen of 24 samples were positive for *Salmonella*, with a mean of 10.9 most probable number (MPN)/100 mL in the positive samples. *Salmonella* were isolated from seven of eight water samples from both Tank 1 and Tank 2, but in only two of eight water samples in Tank 3. The authors concluded that it appears that most bacteria removed from carcasses during scalding are washed off during the early part of scalding in countercurrent systems. This study confirmed that dilution has a beneficial impact on *Salmonella* on poultry.

Cason et al. (2001) showed that using lower scald temperatures in the first scalder does not have an impact on *Salmonella* prevalence on carcasses. Scalding with unheated water in the first tank of a simulated three-tank scalder was tested

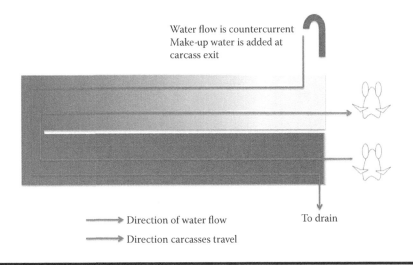

Water flow is countercurrent
Make-up water is added at
carcass exit

To drain

→ Direction of water flow

→ Direction carcasses travel

Figure 13.7   A diagram depicting an ideal countercurrent scalder setup.

to determine whether carcass bacteria, efficiency of feather removal, and cooked breast meat tenderness were affected as compared with carcasses scalded at the same temperature (57°C) in all tanks. The authors found no differences in numbers of aerobic bacteria and *E. coli*, prevalence of *Salmonella*, tenderness of cooked breast meat, or number of feathers left on carcasses. Cason et al. (2001) observed that, in rinses of partially processed carcasses, there were no significant differences between aerobic plate count (APC) or *E. coli* counts in rinses from carcasses that were subjected to normal scalding temperatures or to lower scald temperatures. The lower temperature in the first tank had no effect on carcass microbiology as measured by APC, *E. coli*, and incidence of salmonellae. These data are interesting but are in opposition to data I have observed from commercial processors, where levels of *Salmonella* in the first tank of a three-stage scalder reached greater than 10^5 colony-forming units (CFU)/mL of scalder water. In this particular plant, the first scalder (temperature 40°C) was acting as a *Salmonella* inoculation chamber, increasing *Salmonella* prevalence on carcasses.

An ideal scalder setup is depicted in Figure 13.7.

13.5 Effect of Dilution

By introducing plenty of freshwater into the scalder (at the exit end), a significant portion of the organic material can be removed from the surfaces of the carcass. If this material is allowed to remain on the carcass, it will be transferred into the chiller. If the chiller contains high levels of organic material, then oxidative sanitizers, such as chlorine, will have little effect on bacterial concentrations. Thus,

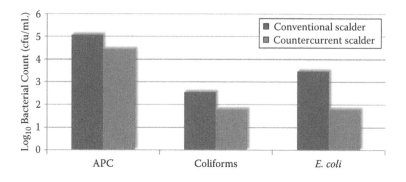

Figure 13.8 Log$_{10}$ **colony-forming units per milliliter of APC, coliforms, and** *E. coli* **on conventionally scalded and countercurrent-scalded chickens. (From Waldroup A, Rathgeber B, and Imel N, 1993, Microbiological aspects of counter current scalding,** *Journal of Applied Poultry Research,* **2, 203–207.)**

maintaining proper flow direction and water flow rate should increase the efficacy of chlorine as it is used later in the process to kill bacteria.

Waldroup et al. (1993) conducted a study to evaluate the effects of dual-stage countercurrent scalding alone on the microbiological condition of broiler chickens. Results of this study are presented in Figure 13.8. Figure 13.9 shows *Salmonella* prevalence on conventionally scalded and countercurrent-scalded chickens (Waldroup et al., 1993). These data demonstrated that countercurrent scalding produces better reduction in bacterial counts than conventional scalding methods. In fact, the study by Waldroup et al. (1993) demonstrated that a properly run countercurrent scalder was able to reduce *Salmonella* prevalence on carcasses by 88.5% without the use of chemicals.

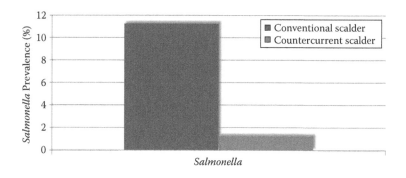

Figure 13.9 *Salmonella* **prevalence on conventionally scalded and counter-current-scalded chickens. (From Waldroup A, Rathgeber B, and Imel N, 1993, Microbiological aspects of counter current scalding,** *Journal of Applied Poultry Research,* **2, 203–207.)**

13.6 Problem with Lowering Scalder Temperatures without Use of Sanitizers

Some companies that use multistage scalders have begun to decrease the temperature in the first scalder tank from the normal temperature range of 53–56°C to 38–42°C. In contrast to the study by Cason et al. (2001), this should not be done because excreta, as it comes off the surface of the birds, will sink to the bottom of the scalder tank. As the shift progresses, increasing excreta may build up in the bottom of the scalder. If the temperature of the scalder is 38–42°C and there are any *Salmonella* in the excreta at the bottom of the tank, the *Salmonella* will begin to multiply rapidly. In fact, this would be near the optimal growth temperature for *Salmonella*. Essentially, at 38°C, the processor is operating the world's most expensive *Salmonella* incubator. In the scalder tank at low temperature, the *Salmonella* cells have all of the components they need to multiply rapidly. Under these conditions, *Salmonella* have the optimum growth temperature, moisture, nutrients, and pH. Thus, the processor will essentially be inoculating bird after bird with *Salmonella* as they go through the scalder.

13.7 Countercurrent Scalding

Most older scalders are like a bath (Figures 13.10 and 13.11), as opposed to having a countercurrent flow. This countercurrent flow has the effect of washing the chickens, much as a fast-moving river would wash dirt from a person better than would a bathtub. However, many poultry companies have difficulty in increasing water flow rate because of municipal water supplier limits.

Figure 13.12 shows the scalder exit of a scalder that is operating correctly. The droplets of water coming off the birds are clear, and the water at the exit is fairly clean.

If the surface of the carcass is contaminated with *Salmonella* in the scalder as a result of the bacteria being transferred from bird to bird, another problem may occur in the next processing step. In the picker, feathers are removed, and the bacteria in the contaminated water from the scalder may be transferred from bird to bird.

13.8 Evaluating Scalder Efficacy

By evaluating the microbiological counts or *Salmonella* prevalence pre- and post-scald, it is possible to determine if the scalder is operating appropriately. Figure 13.13 is an excellent example of an improperly operating scalder. Reasons for this problem may be the following: (1) The temperature of the scalder is far too low to prevent *Salmonella* from growing; (2) not enough fresh makeup water is being added to the scalder; or (3) the water is flowing in the same direction as the carcasses as opposed to a countercurrent flow.

Figure 13.10   A scalder that is not operating in a countercurrent fashion showing filthy scald water and clumped fecal material on carcasses.

Figure 13.11   A scalder that is not operating in a countercurrent fashion showing filthy scald water.

Figure 13.12 Water droplets coming off of the carcasses show that the scalder water at the scalder exit is clean.

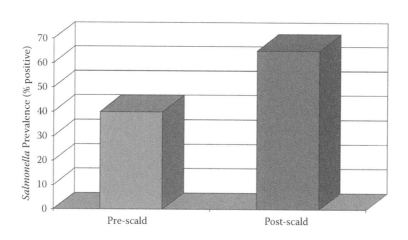

Figure 13.13 *Salmonella* prevalence on carcasses before and after an improperly operating scalder.

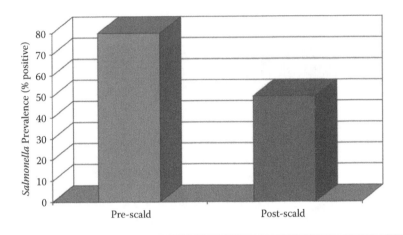

Figure 13.14 *Salmonella* **prevalence on carcasses before and after a properly operating scalder.**

Some suggestions for correction of this problem are as follows:

1. Balance the scalders in terms of freshwater makeup and flow direction (countercurrent).
2. Make sure that the temperature is above 50.5°C in all scald tanks.
3. Add a postbleed, prescald bird brush system to remove any clumps of fecal material from the carcasses prior to scalding.
4. Consider introduction of an approved chemical into the scald tanks.

Once these corrections have been made, conduct additional microbiological evaluations. Figure 13.14 shows a more realistic scenario in a plant that is running the scalder correctly.

The scalder should be considered a positive intervention step in controlling *Salmonella*. If biomapping of the plant indicates that the scalder is significantly contributing to cross contamination of *Salmonella* between carcasses, then steps should be taken to correct the problem.

13.9 Addition of Chemicals to the Scalder and Lowering Temperatures

A survey conducted by the U.S. Poultry and Egg Association (http://www.fsis.usda.gov/PDF/Slides_022406_EKrushinskie.pdf) indicated that, of the poultry companies that add anything to the scalders (which few do), 50% use chlorine and 50% use sodium hypochlorite. Chlorine should not be used in scalders because it is immediately deactivated by the organic load in the scalder and can gas off due to the high temperatures used. Sodium hypochlorite does not have a significant impact on the bacterial levels on carcasses during scalding.

13.9.1 Addition of Acetic, Lactic, and Formic/Proprionic Acids

Cherrington et al. (1992) studied addition of acetic acid, lactic acid, or a commercial preparation of formic and propionic acid (BioAdd) to scalders at a concentration of 0.1% (w/w) at 20°C, 30°C, 40°C, and 50°C in the presence of organic material for activity against *Salmonella kedougou*. BioAdd was the most effective of the solutions evaluated at all temperatures, followed by lactic acid and then acetic acid. The presence of organic material did not significantly affect the antibacterial activity of the acids.

13.9.2 Addition of Sodium Hypochlorite, Acetic Acid, Trisodium Phosphate, and Sodium Metabisulfite

In 1997, Tamblyn et al. conducted experiments, utilizing the skin attachment model (SAM), to determine the bactericidal activity during a simulated scalder (50°C for 2 minutes) application of six potential carcass disinfectants: 20-, 400-, and 800-ppm sodium hypochlorite; 5% acetic acid; 8% trisodium phosphate; and 1% sodium metabisulfite. Efficacies of treatments were determined against populations of *Salmonella typhimurium* that were "loosely" or "firmly" attached to chicken breast skin. Sodium metabisulfite did not reduce populations of *Salmonella*. Sodium hypochlorite at 20 ppm had little activity in the scalder. Trisodium phosphate was similarly effective (reduction by 1.2 to 1.8 $\log_{10}$ CFU per skin sample). Acetic acid was effective in the scalder application and reduced *Salmonella* by 2.0 $\log_{10}$ (Tamblyn et al., 1997). The authors concluded that attachment of *S. typhimurium* to poultry skin apparently increased the ability of the bacteria to resist various disinfectants, and efficacy was influenced by extent of attachment of bacteria to skin and method of disinfectant application.

13.9.3 Addition of Organic Acids

In another study, Tamblyn and Conner (1997) determined the bactericidal activity of organic acids (0.5% and 1%) when combined with various chemical agents (potential transdermal synergists). The SAM was used to determine bactericidal activity of treatments against *Salmonella typhimurium* attached to chicken breast skin. The authors reported that addition of ethanol to all organic acids (0.5%) did not increase their activity and in some cases reduced activity. Addition of Span 20 in the scalder increased activity of all acids against loosely attached cells. Counts were reduced by an additional 0.81 (acetic acid) to 2.35 (tartaric acid) $\log_{10}$ CFU/skin. Span 20 also increased activity of citric acid against firmly attached cells by 1.63 $\log_{10}$ CFU/skin in the scalder. When added to a 1% concentration of the acids, sodium lauryl sulfate caused the greatest increase in antimicrobial activity (Tamblyn and Conner, 1997).

13.9.4 Addition of Sulfuric Acid, Ammonium Sulfate, and Copper Sulfate

Russell (2008) conducted a study to evaluate the effect of an acidic, copper sulfate-based commercial sanitizer on total aerobic bacteria (APC) and *Escherichia coli* counts and *Salmonella* prevalence on broiler chicken carcasses when applied during scalding or scalding and postpick dipping. When applied during scalding in a commercial processing plant, APC and *E. coli* counts were significantly ($p < 0.05$) reduced on all days of sampling. The average $\log_{10}$ reduction overall was 3.80 and 3.05 for APC and *E. coli*, respectively. *Salmonella* prevalence was reduced by an average of 30%. For carcasses that were scalded, picked, and dipped postpick using this sanitizer, APC were significantly ($p < 0.05$) reduced on all days of sampling by an average of 1.19 $\log_{10}$. *Escherichia coli* counts were reduced on all but 2 days of sampling for carcasses scalded, picked, and dipped in this sanitizer (Russell, 2008).

13.9.5 Addition of Sodium Hydroxide

In 2008, McKee et al. evaluated the efficacy of a scalder additive made of 1% sodium hydroxide (RP Scald) at pH 11.0 on *Salmonella typhimurium* levels on inoculated poultry carcasses. This study showed that inoculated broilers that were scalded at higher temperatures together with the 1% NaOH had the lowest *Salmonella* recovery. McKee et al. (2008) concluded that 1% NaOH may be effective in reducing *Salmonella typhimurium* on broiler carcasses in poultry scalder applications, particularly when hard scald temperatures are used.

13.9.6 Effect of a Flow-Through Electrical Treatment System

Efforts have been made to reduce *Salmonella* in scalder water without the use of chemicals. Wolfe et al. (1999) studied a flow-through electrical treatment system as a method to eliminate *Salmonella typhimurium* in poultry scalder water. The flow-through treatment system consisted of an inlet tank, a treatment cylinder, and an outlet tank. A DC power supply provided a 2-ampere electrical current to the poultry scalder water flowing through the treatment cylinder at a rate of 5 L/minute. Residence time in the cylinder was 1.82 minutes. Using this technique, all *Salmonella* were eliminated from the treated poultry scalder water, while the untreated scalder water contained large numbers of *Salmonella* (Wolfe et al., 1999).

References

Abu-Ruwaida A S, Sawaya W N, Dashti B H, Murad M, and Al-Othman H A (1994), Microbiological quality of broilers during processing in a modern commercial slaughterhouse in Kuwait, *Journal of Food Protection*, 57, 887–892.

Bryan F L, and Doyle M P (1995), Health risks and consequences of *Salmonella* and *Campylobacter jejuni* in raw poultry, *Journal of Food Protection*, 58, 326–344.

Cason J A, Hinton A, Jr, and Buhr R J (2001), Unheated water in the first tank of a three-tank broiler scalder, *Poultry Science*, 80, 1643–1646.

Cason J A, Hinton A, Jr, and Ingram K D (2000), Coliform, *Escherichia coli*, and salmonellae concentrations in a multiple-tank, counterflow poultry scalder, *Journal of Food Protection*, 63, 1184–1188.

Cason J A, Whittemore A D, and Shackelford A D (1999), Aerobic bacteria and solids in a three-tank, two-pass, counterflow scalder, *Poultry Science*, 78, 144–147.

Cherrington C A, Allen V, and Hinton M (1992), The influence of temperature and organic matter on the bactericidal activity of short-chain organic acids on salmonellas, *Journal of Applied Bacteriology*, 72, 500–503.

Kim J W, Slavik M F, Griffis C L, and Walker J T (1993), Attachment of *Salmonella* Typhimurium to skins of chicken scalded at various temperatures, *Journal of Food Protection*, 56, 661–665, 671.

McKee S R, Bilgili S F, and Townsend J C (2008), Use of a scald additive to reduce levels of *Salmonella* Typhimurium during poultry processing, *Poultry Science*, 87, 1672–1677.

Russell S M (2008), The effect of an acidic, copper sulfate-based commercial sanitizer on indicator, pathogenic, and spoilage bacteria associated with broiler chicken carcasses when applied at various intervention points during poultry processing, *Poultry Science*, 87, 1435–1440.

Tamblyn K C, and Conner D E (1997), Bactericidal activity of organic acids in combination with transdermal compounds against *Salmonella* Typhimurium attached to broiler skin, *Food Microbiology*, 14, 477–484.

Tamblyn K C, Conner D E, and Bilgili S F (1997), Utilization of the skin attachment model to determine the antibacterial efficacy of potential carcass treatments, *Poultry Science*, 76, 1318–1323.

Waldroup A, Rathgeber B, and Imel N (1993), Microbiological aspects of counter current scalding, *Journal of Applied Poultry Research*, 2, 203–207.

Wolfe R E, Griffis C L, Li Y, Slavik M F, and Walker J T (1999), Effect of electrical treatment on *Salmonella* Typhimurium in poultry scalder water using a flow-through treatment system, *Applied Engineering in Agriculture*, 15, 535–537.

Chapter 14

Improving Processing Yield and Lowering *Salmonella* during Scalding without Added Expense

14.1 Introduction

For years, poultry companies have known that they can significantly improve finished whole-ready-to-cook without giblets (WOG) yield by lowering scalder temperatures. However, every time companies attempt to lower scalder temperatures, two problems routinely occur: (1) *Salmonella* begins to multiply in the scalder, and (2) carcasses do not pick well. Many companies that attempt this technique have difficulty meeting the hazard analysis and critical control point (HACCP)/ *Salmonella* pathogen reduction rule of the Food Safety Inspection Service (FSIS) of the USDA (U.S. Department of Agriculture). After carefully assessing these plants, the only mistake that was being made was that they were running the scalders at temperatures below 50.5°C. As mentioned, the maximum growth temperature for *Salmonella* is 45°C. The research literature indicates that, to preclude the growth of microorganisms, they should be kept at a temperature that is at least 5°C above their maximum growth temperature. Thus, to prevent any multiplication of *Salmonella* in scalders, the temperature should never be lower than 50.5°C. After evaluating scalders in processing plants that were running at 46°C, 100,000 *Salmonella* cells were found in each milliliter of scalder water. Thus, every chicken that was being processed was being inoculated with hundreds of thousands of *Salmonella* cells. Keep in mind that only a single *Salmonella* cell results in a positive carcass postchill and may be detected as a positive carcass by USDA-FSIS. This makes the job of the intervention strategy that is employed downstream, such as the online reprocessing (OLR) system or the chiller, that much more difficult.

14.2 Preventing *Salmonella* Growth While Using Reduced Scalder Temperatures

Russell (2007) wrote an article detailing how this problem can be overcome. This problem may be avoided by introducing a sanitizer into the scalder. It seems easy to put a chemical sanitizer into the scalder, but in actual practice, it is difficult. For this reason, few poultry companies add anything to their scalders to disinfect them. The reasons are twofold: (1) They are not used to spending money on scalder disinfectants and are reticent to spend the extra money, and (2) few chemicals have been shown to have any efficacy. I have observed numerous instances when a plant manager agrees to add a disinfectant to the scalder. The company begins to see excellent *Salmonella* reduction on postchill carcasses; however, when an accountant in the main office sees the added expense for this disinfectant, the accountant immediately cuts the expenditure. This is because there is, in some cases, a serious lack of communication between the people who actually produce the chicken and those who work in the main office paying bills. This problem must be addressed first by educating those responsible for expenditures why the added expense is necessary.

14.2.1 Using Oxidant Sanitizers

Chlorine or other oxidant-based sanitizers should not be used in scalders because they are immediately deactivated by the organic load in the scalder and can gas off. Sodium hypochlorite (a basic compound that raises the pH dramatically) does not have a significant impact on the bacterial levels on carcasses during scalding.

14.2.2 Acid Sanitizers

A benefit to adding acidic disinfectant chemicals such as inorganic acids, organic acids, peracetic acid, or mixtures thereof to the scalder is that the scalder temperature may then be lowered without the concern that *Salmonella* will begin to grow. Likewise, acids greatly improve picking by denaturing the protein that holds the feather in the follicle. In studies in very large-scale processing plants, scalder pH was lowered to 2.0, 3.0, and 4.0 in scald Tanks 3, 2, and 1, respectively. No *Salmonella* cells were detected in any of the three scald tanks. Also, by doing this, all of the carcasses appeared to be completely picked on exiting the first picker. Thus, scalder temperatures could be greatly reduced while avoiding the two problems with low scald temperature noted. Moreover, using a mixture of sulfuric acid, ammonium sulfate, and copper sulfate in the scalder in these studies resulted in a dramatic reduction in aerobic plate counts (APCs) and *E. coli* counts on chicken carcasses, and this reduction was maintained throughout the rest of the process.

14.3 How Does Reducing Scalder Temperature Improve Yield?

What is the logical reason why lowering scalder temperatures improves yield? Chicken fat is very unsaturated, which means that it becomes liquefied at lower temperatures than saturated fat like beef fat (tallow) or pork fat (lard). This means that even with only 2 minutes of exposure to a scalder at a temperature of 54.5°C and above, the fat under the skin becomes liquefied (see Figure 14.1) and will drain out from under the skin as the carcasses move along the line suspended upside down and

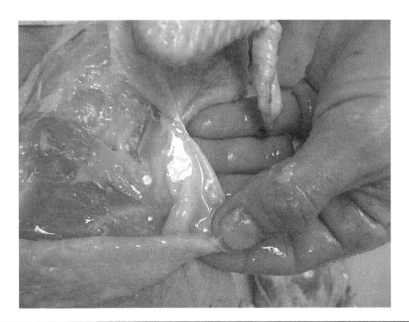

Figure 14.1   Liquefied fat under chicken skin immediately postscald and pick.

Figure 14.2 Fat that has been released into the chiller due to excessive scalder temperatures.

especially when the carcasses are placed into the chiller, where there is tremendous agitation (see Figure 14.2). This results in yield loss in the form of fat loss. This is different from yield loss or gain because of water pickup because much of the water pickup will be lost later due to weepage and is viewed negatively by consumers.

Many companies have noticed that when they switched from trisodium phosphate OLR applications, they encountered a significant yield loss, generally on the order of 1–2%. This is a huge loss and causes a great deal of consternation. However, the yield gain because of water pickup in the chiller is a problem. The trisodium phosphate shifts the pH upward, thus making the pH of the meat slightly higher and away from its isoelectric point. This makes the water-holding capacity of the meat increase greatly. This is why phosphate is almost always used in marinades to make the meat hold the marinade that is introduced. The problem is that the meat will eventually purge some of this water into the package later, and customers do not like liquids spilling out of their packages. Therefore, yield increase due to water weight gain is not the best option. It is much more advisable to maintain the fat under the skin of the chicken. The benefits are as follows:

1. More subskin fat equates to more WOG yield.
2. Less fat is expressed onto equipment, reducing the ability of *Salmonella* to be encased in this fat and be protected from chemical sanitizers and contaminate carcasses as they touch the equipment.

3. Less fat comes out from under the skin in the chiller, which reduces the organic load of the chiller, allowing chlorine and other oxidants to do a much better job.
4. Less fat must be removed from the overflow water of the chiller by the waste-water treatment system.

14.4 How to Achieve a Yield Increase

How much yield can you gain by lowering scalder temperatures? A study was conducted by Russell (2007) in two very large poultry-processing facilities. In the first approach, the temperature of all three scald tanks was lowered by 5°C. The birds did not pick well using this approach. Instead, ramping up the scald temperatures and gradually increasing temperature from the initial scalder entrance to the scalder exit will result in excellent picking while improving yield. In the first plant, which processes over 320,000 carcasses per day, excellent pick was maintained while keeping the three scald tanks at 43.3°C, 45.6°C, and 58.9°C for Tanks 1, 2, and 3, respectively. The plant originally ran the scalder at 55.6°C, 56.7°C, and 57.8°C for Tanks 1, 2, and 3, respectively (see Figure 14.3).

Plant managers have expressed the opinion that reducing the temperature of the scalders only a few degrees results in observable yield increases. In one study with a large processor, a 2% yield increase was achieved. Yield was measured by weighing 100 live chickens, tagging them, and reweighing them after chilling. By comparing the WOG yield to historical values achieved for yield, a 2% increase was noted. In another plant, unfortunately a one-stage scalder was being used. This makes it difficult to adjust the temperatures such that they ramp up rapidly. In this plant,

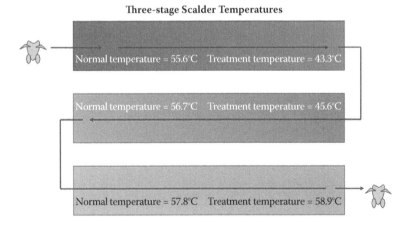

Three-stage Scalder Temperatures

Normal temperature = 55.6°C   Treatment temperature = 43.3°C

Normal temperature = 56.7°C   Treatment temperature = 45.6°C

Normal temperature = 57.8°C   Treatment temperature = 58.9°C

Figure 14.3   Scalder temperatures that are normally used in the three tanks and the temperatures used in this study.

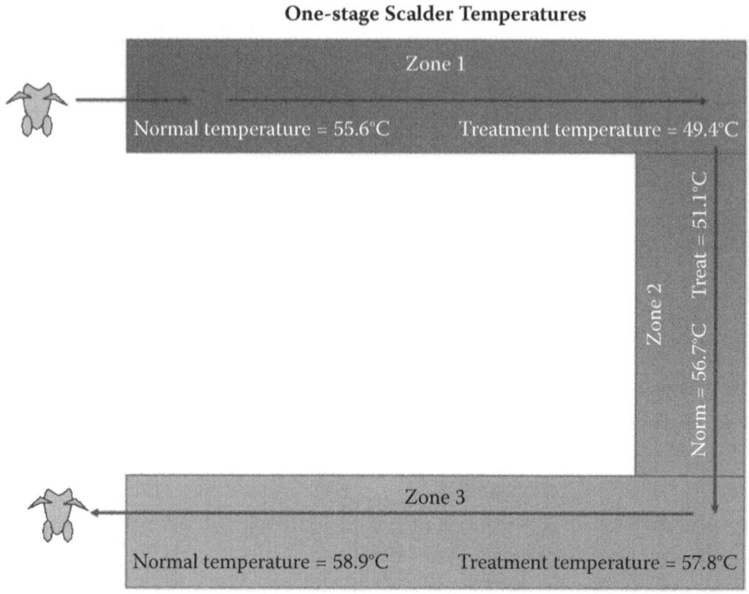

One-stage Scalder Temperatures

Figure 14.4 Scalder temperatures that are normally used in the single-stage scalder tank and the temperatures used in this study.

the steam inputs into the scald tank were carefully adjusted to improve yield (see Figure 14.4).

The scalders were run at normal temperatures for the control group at 55.6°C, 56.7°C, and 57.8°C for Zones 1, 2, and 3, respectively. For the test group, sulfuric acid, ammonium sulfate, and copper sulfate were added to the scalder (pH 2.0), and the zones were operated at 49.4°C, 51.1°C, and 58.9°C for Zones 1, 2, and 3, respectively. The carcasses were evaluated by counting feathers remaining after pick and found to be equivalent for the control and treated groups of chickens. Live birds (50) were tagged, weighed, then slaughtered and processed, collected after chilling, and reweighed for both the control and treated groups. This study was conducted over 4 days of processing.

14.5 How Much Yield Can Be Maintained?

The results (Russell, 2007) were very promising. Over the 3 days of testing, the processing yield for the control group was 76.19%. However, for carcasses scalded at lower temperatures, the yield was 76.94% (see Figure 14.5). This means that an average yield increase of 0.75% was achieved (see Figure 14.6). This yield increase was achieved under the worst-possible conditions. Using a single-stage scalder made it hard to control the temperature zones and maintain picking quality. It has been

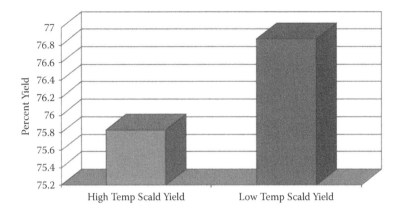

Figure 14.5 Percentage yield for chicken carcasses scalded using normal or low-temperature scalding in the plant with a one-stage scalder.

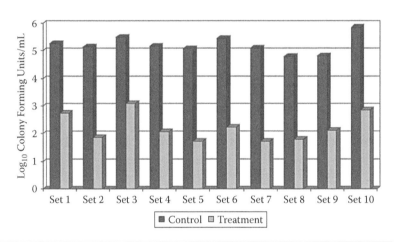

Figure 14.6 APCs on postscald broiler carcasses for carcasses scalded in tap water versus an acid sanitizer.

observed that much greater yield increases would be achieved when a three-stage scalder is used instead of the one scalder used in this study because in three-stage scalders, the temperatures may be lowered to a much greater degree, as described previously.

14.6 Savings Calculation

If an average poultry processor produces 200,000 to 250,000 chickens per day at an average weight of 3 lb, the yield increase observed in this study equates to 600,000 to 750,000 lb of meat. If a figure of $0.70 cents/lb of meat is used, then the plant would be producing $420,000 to $525,000 worth of product per day. Thus, using these figures, a 0.75% yield increase equates to $3,150 to $3,937 a day in savings (or $819,000 to $1,023,620 per year) even when only adjusting the scalder tempera-tures down a moderate degree. Also, a minimum of $500 per day would be saved in energy costs, resulting in a total cost savings of $130,000 per year. Overall, using a proper acid sanitizer in the scalder has been demonstrated to

1. Make carcasses much easier to pick
2. Decrease bacterial growth and lower bacterial numbers on carcasses, including *Salmonella*
3. Decrease cross contamination of *Salmonella* from carcass to carcass
4. Reduce overscald striping of breasts
5. Reduce the cost of energy required to heat the scalder water
6. Reduce the amount of subskin fat cook-off, leading to higher yields
7. Decrease the amount of fat on processing equipment
8. Decrease the amount of fat in the chiller, allowing chlorine or other chemis-tries to do a much better job

Assuming that an acid sanitizer costs a plant $2,500 per day to use (this would be a very high estimate), the cost benefit would be approximately $650 to $1,437 per day (or $169,000 to $373,620 per year). Keep in mind that when the purchas-ing employee in the main office sees this expense, he or she is likely to have a nega-tive reaction because no one has explained the financial benefit. There is another cost that is rarely considered by processing plant personnel. The cost to remediate *Salmonella* in the field can be high. Although this cost does not show up on the bottom line for a processing plant manager, it is ultimately incurred by the com-pany. For example, *Salmonella* vaccines used on breeder chickens and on broiler chickens may be expensive (0.6 to 0.9 cents per bird). Thus, for those companies using vaccines on broilers alone, that would be a daily savings of $1,500 to $2,250 for the larger-scale plant (250,000 birds per day) for a total savings of $390,000 to $585,000 per year. Acids used in water systems during feed withdrawal to eliminate *Salmonella* picked up during feed withdrawal when birds consume feces is another cost in the field. The average cost is $50 per house times 10 houses processed per day, which equates to a $500-per-day savings or a savings of $130,000 per year. By eliminating the need for these interventions in the field, the company saves addi-tional money due to this approach. Field interventions can be eliminated because, even in companies that routinely have incoming *Salmonella* contamination rates of 100%, the number of positive carcasses postchill using scalder, picker, and postpick spray disinfection is usually less than 5% postchill. Thus, there is no real need to use preharvest intervention when multiple hurdles are used properly, effectively eliminating their cost.

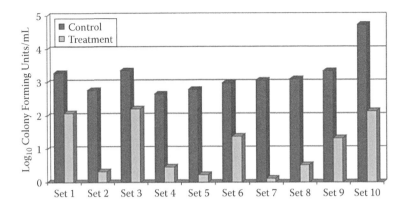

Figure 14.7 *Escherichia coli* **counts on postscald broiler carcasses for carcasses scalded in tap water versus an acid santizer.**

14.7 Bacterial Reductions Achieved Using an Acid Sanitizer

Russell (2008) conducted a study using sulfuric acid, ammonium sulfate, and copper sulfate in a commercial scalding system. Figures 14.6 and 14.7 show the reduction in APCs and *E. coli* after scalding (control scalder used tap water and treated scalder on an adjacent identical line was treated at a pH of 2, 3, and 4 in Tanks 3, 2, and 1, respectively) when using a low pH.

Using a chemical sanitizer in the scalder in these studies had a dramatic impact on APCs and *E. coli* counts on chicken carcasses (Russell, 2008). Using acid sanitizers has been shown to have numerous benefits, and they are a cost-effective means of reducing *Salmonella*.

References

Russell S M (2007), Solving the yield/pathogen reduction dilemma, *PoultryUSA*, October, 30–32.

Russell S M (2008), The effect of an acidic, copper sulfate-based commercial sanitizer on indicator, pathogenic, and spoilage bacteria associated with broiler chicken carcasses when applied at various intervention points during poultry processing, *Poultry Science*, 87, 1435–1440.

Chapter 15

The Effect of Picking (Defeathering) on *Salmonella* Levels on Carcasses

15.1 Introduction

A number of studies have clearly demonstrated the negative impact of defeathering (picking) on *Salmonella* levels on broiler carcasses (Figure 15.1). The defeathering process, which consists of scalding, followed by mechanical feather removal, is a site of significant microbial cross contamination (Ono and Yamamoto, 1999; Nde et al., 2006). Potential mechanisms of cross contamination with *Salmonella* during defeathering include aerosols (Allen et al., 2003a, 2003b), direct contact between contaminated and uncontaminated carcasses (Mulder et al., 1978), and the action of the fingers of the picker machines (Clouser et al., 1995a, 1995b; Berrang et al., 2001; Allen et al., 2003a, 2003b). Supporting the idea that aerosols can be a source of transmission, Lues et al. (2007) observed that the highest counts of microorganisms were recorded in the initial stages of processing, comprising the receiving-killing and defeathering areas.

Many of the studies concerning cross contamination during picking have focused on the change in *Salmonella* prevalence that occurs before and after defeathering

Figure 15.1   Chicken picking system.

and the dissemination pattern of an indicator organism from an artificially contaminated bird to other birds during processing (Clouser et al., 1995a, 1995b; Allen et al., 2003a, 2003b). For example, Abu-Ruwaida et al. (1994) found that microbial levels varied during processing, but the highest levels were detected after scalding and defeathering.

15.2 Cross Contamination with *Salmonella* during Picking as Determined by Subtyping

Nde et al. (2007) reported that no study has examined the relationship between *Salmonella* subtypes isolated before and after defeathering. Molecular typing is increasingly being used to complement conventional methodologies to elucidate bacterial transmission routes (De Cesare et al., 2001; Sander et al., 2001). Several studies have used molecular typing to understand the transmission of *Salmonella* within poultry production (De Cesare et al., 2001; Bailey et al., 2002; Liebana et al., 2002; Crespo et al., 2004). Nde et al. (2007) sampled the breast feathers of turkeys before defeathering and the exposed breast skin of carcasses after defeathering to determine the extent of cross contamination occurring during defeathering. Serotyping and pulsed-field gel electrophoresis (PFGE) were used to more specifically determine the relationship between *Salmonella* isolates obtained pre- and post-defeathering. Figure 15.2 shows the prevalence of *Salmonella* on turkeys before and after feathering.

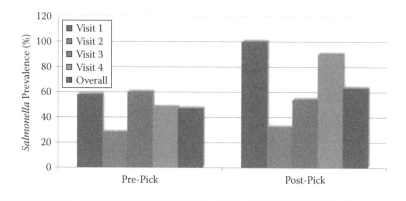

Figure 15.2 *Salmonella* **prevalence on turkeys before and after picking. (From Nde C W, Mcevoy J M, Sherwood J S, and Logue C M (2007), Cross-contamination of turkey carcasses by** *Salmonella* **species during defeathering,** *Poultry Science,* **86, 162–167.)**

This study showed that *Salmonella* was transferred from live turkeys to carcass skin during defeathering (Nde et al., 2007). In all cases except Visit 3, more turkeys were positive for *Salmonella* after picking than were positive before picking. Overall, picking increased *Salmonella* prevalence by 16%. The authors reported that contamination on the external surface of a bird is as much a source of *Salmonella* as expelled feces. Other authors have studied chicken and turkey carcass defeathering operations (Lillard, 1989, 1990; Clouser et al., 1995a, 1995b) and found that *Salmonella* contamination after defeathering was observed to increase or remain the same. Nde et al. (2007) also found that the same serotypes found on turkeys before picking were found on the turkey carcasses after picking. Reich et al. (2008), studying *Campylobacter,* also found that the numbers of *Campylobacter* were highest in carcasses after scalding/defeathering (mean 5.9 $\log_{10}$ colony-forming units [CFU]/carcass).

References

Abu-Ruwaida A S, Sawaya W N, Dashti B H, Murad M, and Al-Othman H A (1994), Microbiological quality of broilers during processing in a modern commercial slaughterhouse in Kuwait, *Journal of Food Protection*, 57, 887–892.

Allen V M, Hinton M H, Tinker D B, Gibson C, Mead G C, and Wathes C M (2003a), Microbial cross-contamination by airborne dispersion and contagion during defeathering of poultry, *British Poultry Science*, 44, 567–576.

Allen V M, Tinker D B, Hinton M H, and Wathes C M (2003b), Dispersal of microorganisms in commercial defeathering systems, *British Poultry Science*, 44, 53–59.

Bailey J S, Cox N A, Craven S E, and Cosby D E (2002), Serotype tracking of *Salmonella* through integrated broiler chicken operations, *Journal of Food Protection*, 65, 742–745.

Berrang M E, Buhr R J, Cason J A, and Dickens J A (2001), Broiler carcass contamination with *Campylobacter* from feces during defeathering, *Journal of Food Protection*, 64, 2063–2066.

Clouser C S, Doores S, Mast M G, and Knabel J S (1995a), The role of defeathering in the contamination of turkey skin by *Salmonella* species and *Listeria monocytogenes*, *Poultry Science*, 74, 723–731.

Clouser C S, Knabel J S, Mast M G, and Doores S (1995b), Effect of type of defeathering system on *Salmonella* cross-contamination during commercial processing, *Poultry Science*, 74, 732–741.

Crespo R, Jeffrey J S, Chin R P, Senties-Cue G, and Shivaprasad H L (2004), Phenotypic and genotypic characterization of *Salmonella arizonae* from an integrated turkey operation, *Avian Diseases*, 48, 344–350.

De Cesare A, Manfreda G, Dambauh T R, Guerzoni M E, and Franchini A (2001), Automated ribotyping and random amplified polymorphic DNA analysis for molecular typing of *Salmonella* Enteritidis and *Salmonella* Typhimurium isolated in Italy, *Journal of Applied Microbiology*, 91, 780–785.

Liebana E, Crowley C J, Garcia-Migura L, Breslin M F, Corry J E L, Allen V M, and Davies R H (2002), Use of molecular fingerprinting to assist the understanding of the epidemiology of *Salmonella* contamination within broiler production, *British Poultry Science*, 43, 38–46.

Lillard H S (1989), Incidence and recovery of salmonellae and other bacteria from commercially processed poultry carcasses at selected pre-and post-evisceration steps, *Journal of Food Protection*, 52, 88–91.

Lillard H S (1990), The impact of commercial processing procedures on the bacterial contamination and cross contamination of broiler carcasses, *Journal of Food Protection*, 53, 202–204.

Lues J F R, Theron M M, Venter P, and Rasephei M H R (2007), Microbial composition in bioaerosols of a high-throughput chicken-slaughtering facility, *Poultry Science*, 86, 142–149.

Mulder R W A W, Dorresteijn L W J, and Van Der Broek J (1978), Cross-contamination during the scalding and plucking of broilers, *British Poultry Science*, 19, 61–70.

Nde C W, Mcevoy J M, Sherwood J S, and Logue C M (2007), Cross-contamination of turkey carcasses by *Salmonella* species during defeathering, *Poultry Science*, 86, 162–167.

Nde C W, Sherwood J S, Doetkott C, and Logue C M (2006), Prevalence and molecular profiles of *Salmonella* collected at a commercial turkey processing plant, *Journal of Food Protection*, 69, 1794–1801.

Ono K, and Yamamoto K (1999), Contamination of meat with *Campylobacter jejuni* in Saitama, Japan, *International Journal of Food Microbiology*, 47, 211–219.

Reich F, Klein G, Haunhorst E, and Atanassova V (2008), The effects of *Campylobacter* numbers in caeca on the contamination of broiler carcasses with *Campylobacter*, *International Journal of Food Microbiology*, 127, 116–120.

Sander J, Hudson C R, Dufour-Zavala L, Waltman W D, Lobsinger C, Thayer S G, Otalora R, and Maurer J J (2001), Dynamics of *Salmonella* contamination in a commercial quail operation, *Avian Diseases*, 45, 1044–1049.

Chapter 16

Methods for Controlling *Salmonella* Levels on Carcasses during Picking

16.1 Introduction

Few studies have examined methods for reducing *Salmonella* during picking. Dickens and Whittemore (1997) evaluated the use of acetic acid or H_2O_2 spray during defeathering. This may be accomplished by placing the chemical into the picker rails (Figure 16.1). The authors reported that aerobic plate counts (APCs) were significantly reduced when using 1% acetic acid in comparison to the water control (3.93 vs. 4.53 $\log_{10}$ colony-forming units [CFU]/mL). However, the addition of 0.5%, 1%, or 1.5% H_2O_2 to spray waters had no effect on microbiological quality of the carcasses when compared to the water control.

16.2 Effect of Intestinal Contents Released during Picking

Little work on *Salmonella* has been done regarding the release of intestinal contents during picking. However, Berrang et al. (2006b) stated that *Campylobacter* numbers on broiler carcasses during defeathering can increase because of leakage of contaminated gut contents in the picker. The authors attempted to reduce

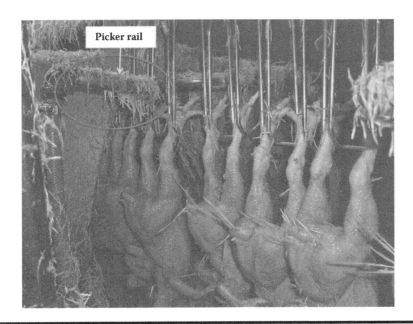

Figure 16.1 The picker rail system in a chicken defeathering system where chemicals may be added.

Campylobacter numbers on chickens during picking by placing food-grade organic acid into the cloaca of chickens before picking. Berrang et al. (2006b) reported that placement of food-grade organic acids in the cloaca of broiler carcasses may be useful as a means to lessen the impact of automated defeathering on the microbiological quality of carcasses during processing. In another study, Berrang et al. (2006a) reported that *Campylobacter* numbers recovered from broiler carcass skin samples increased during automated feather removal. The authors placed vinegar into the colons of the chickens prior to scalding. Carcasses were then scalded, and *Campylobacter* numbers were determined on breast skin before and after passage through a commercial-style picker. *Campylobacter* numbers recovered from the breast skin of untreated control carcasses increased during feather removal from 1.3 $\log_{10}$ CFU per sample prior to defeathering to 4.2 $\log_{10}$ CFU afterward. Berrang et al. (2006a) found that *Campylobacter* numbers recovered from the breast skin of carcasses treated with vinegar also increased during defeathering, but to a significantly lesser extent. Treated carcasses experienced only a 1.0 $\log_{10}$ increase from 1.6 $\log_{10}$ CFU per sample before feather removal to 2.6 $\log_{10}$ CFU per sample afterward. Although this method may not be a practical approach, the authors concluded that application of an effective food-grade antimicrobial in the colon prior to scald may limit the increase in *Campylobacter* contamination of broiler carcasses during defeathering. It is possible that this unique approach may also be used for *Salmonella* reduction.

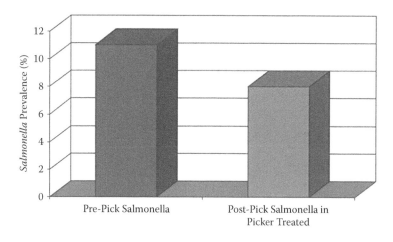

Figure 16.2 *Salmonella* **prevalence on carcasses pre- and postpick in a picker system treated with an acidic sanitizer.**

16.3 Effect of Acidic Sanitizers to Picker Rails

I (unpublished data, 2006) conducted a study in which a blend of sulfuric acid, ammonium sulfate, and copper sulfate was applied to the picker rails of a commercial poultry picking system, and directly adjacent to this picker system, tap water was applied in an identical picker. A total of 80 treated animals and 80 untreated controls were evaluated. The data are presented in Figure 16.2.

Instead of increasing during picking, *Salmonella* prevalence on broiler carcasses was reduced by 3%. This study demonstrated that cross contamination of broilers could be controlled using the sulfuric acid blend described.

References

Berrang M E, Hinton A, Jr, and Smith D P (2006a), Application of distilled white vinegar in the cloaca to counter the increase in *Campylobacter* numbers on broiler skin during feather removal, *Journal of Food Protection*, 69, 425–427.

Berrang M E, Smith D P, and Hinton A, Jr (2006b), Organic acids placed into the cloaca to reduce *Campylobacter* contamination of broiler skin during defeathering, *Journal of Applied Poultry Research*, 15, 287–291.

Dickens J A, and Whittemore A D (1997), Effects of acetic acid and hydrogen peroxide application during defeathering on the microbiological quality of broiler carcasses prior to evisceration, *Poultry Science*, 76, 657–660.

Chapter 17

Reducing Pathogens during Evisceration

17.1 Introduction

Smith et al. (2007) reported that the Food Safety Inspection Service (FSIS) of the U.S. Department of Agriculture (USDA) mandated a zero-tolerance policy for fecal material on poultry carcasses prior to entering the chiller (USDA, 2005). These actions demonstrated that the FSIS is determined to reduce fecal contamination and associated numbers of pathogens on processed poultry. Fecal material or ingesta, and bacteria associated with these contaminants, may be introduced to the broiler carcass during processing (Smith et al., 2007). Damage to the intestines may occur during the evisceration process, leading to carcass contamination (Sams, 2001). Russell (2003) reported that intestines cut during the evisceration process ranged from 2% to 34% of broilers evaluated in one processing plant. At a commercial processing plant, 25% of crops collected at the cropper machine were observed to have been damaged (Hargis et al., 1995). Manual crop removal (which would presumably be more gentle than mechanical cropping) had an overall average of 22% ruptured broiler crops (Buhr and Dickens, 2001). Thus, damage to viscera and crops during evisceration may result in the spread of *Salmonella* to carcasses.

It has been estimated that between 1% and 5% of all broiler chickens produced in the United States must be reprocessed due to disease or contamination of the carcass with fecal material or ingesta. Reprocessing requires additional labor and, if done incorrectly, introduces the opportunity for bacteria to multiply on the carcass while carcasses are taken off-line to be reprocessed. The core temperature of the bird during reprocessing is approximately 39°C, and the reprocessing area is generally room temperature. Therefore, in a short period of time, if carcasses are left under these conditions, pathogenic bacteria may multiply to dangerous levels.

17.2 Effect of Evisceration on Microbial Numbers

In 1994, Abu-Ruwaida et al. reported that bacterial levels did not change during vent opening or evisceration. A study by Jimenez et al. (2002) compared the prevalence of *Salmonella* on chicken carcasses with and without visible fecal contamination during commercial slaughter. For carcasses that were uncontaminated with feces after evisceration, 20% were contaminated with *Salmonella,* and 20.8% of the carcasses that were visibly contaminated were positive. No significant differences in *Salmonella* prevalence on fecally contaminated versus uncontaminated carcasses were observed (Jimenez et al., 2002).

More recent studies on contamination during evisceration were conducted on fecal indicator bacteria and *Campylobacter,* as opposed to *Salmonella.* Berrang et al. (2004) stated that intestinal contents may contaminate broiler carcasses during processing. The authors conducted a study to determine the effect of various levels of intestinal contents on the numbers of *Campylobacter* detected in broiler carcass rinse samples. Berrang et al. (2004) discovered that carcass halves contaminated with only 5 mg of cecal contents had an average of 3.3 log CFU of *Campylobacter* per milliliter of rinse, while corresponding uncontaminated carcass halves had 2.6 log CFU of *Campylobacter* per milliliter of rinse. These data indicate that even small (5-mg) amounts of cecal contents can cause a significant increase in the numbers of *Campylobacter* on eviscerated broiler carcasses. Therefore, it is important to keep such contamination to a minimum during processing (Berrang et al., 2004), and this principle applies to *Salmonella* as well.

Smith et al. (2007) studied the effect of external or internal fecal contamination on the numbers and incidence of coliforms, *Escherichia coli,* and *Campylobacter* after evisceration and passage through an inside-outside bird washer (IOBW). The authors found that external contamination resulted in higher numbers of bacteria after carcass washing, but carcasses with internal contamination still had higher numbers of bacteria after washing than carcasses without applied contamination (Figure 17.1).

17.3 Effect of Evisceration Equipment on Fecal Contamination

Russell and Walker (1996) compared two evisceration systems with regard to fecal contamination. A conventional Streamlined Inspection System (SIS) using

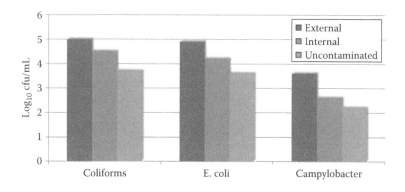

Figure 17.1 Log_{10} colony-forming units per milliliter (CFU/mL) of coliforms, *E. coli*, and *Campylobacter* on uncontaminated chicken carcasses or carcasses contaminated on the inside or outside. (From Smith D P, Northcutt J K, Cason J A, Hinton A, Jr, Buhr R J, and Ingram K D, 2007, Effect of external or internal fecal contamination on numbers of bacteria on prechilled broiler carcasses, *Poultry Science*, 86, 1241–1244.)

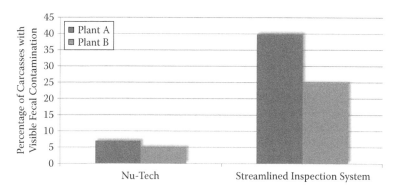

Figure 17.2 Percentage of chicken carcasses with visible fecal contamination for carcasses eviscerated using the Stork Nu-Tech system versus a conventional Streamlined Inspection System.

a traditional spoon eviscerator that removes the viscera and hangs it on the back of the carcass was compared to the Stork Nu-Tech System (Stork-Gamco, Inc., Gainesville, GA), which removes the visceral package during evisceration and the viscera are presented to the inspector just adjacent to the carcass from which it originated. Using the Nu-Tech System, the viscera are never allowed to contact the outside of the carcass. Results are presented in Figure 17.2.

These results clearly demonstrate that the Nu-Tech method of evisceration produces fewer visibly contaminated carcasses than does the traditional SIS evisceration system (Russell and Walker, 1996).

References

Abu-Ruwaida A S, Sawaya W N, Dashti B H, Murad M, and Al-Othman H A (1994), Microbiological quality of broilers during processing in a modern commercial slaughterhouse in Kuwait, *Journal of Food Protection*, 57, 887–892.

Berrang M E, Smith D P, Windham W R, and Feldner P W (2004), Effect of intestinal content contamination on broiler carcass *Campylobacter* counts, *Journal of Food Protection*, 67, 235–238.

Buhr R J, and Dickens J A (2001), Crop extraction load and efficiency of crop removal during manual evisceration of broilers: I. Evaluation of stunning voltage and method of bleeding, *Journal of Applied Poultry Research*, 10, 71–78.

Hargis B M, Caldwell D J, Brewer R L, Corrier D E, and Deloach J R (1995), Evaluation of the chicken crop as a source of *Salmonella* contamination for broiler carcasses, *Poultry Science*, 74, 1548–1552.

Jimenez S M, Salsi M S, Tiburzi M C, and Pirovani M E A (2002), Comparison between broiler chicken carcasses with and without visible faecal contamination during the slaughtering process on hazard identification of *Salmonella* spp., *Journal of Applied Microbiology*, 93, 593–598.

Russell S M (2003), The effect of air sacculitis on bird weights, uniformity, fecal contamination, processing errors, and populations of *Campylobacter* spp and *Escherichia coli*, *Poultry Science*, 82, 1326–1331.

Russell S M, and Walker J M (1996), The effect of evisceration on visible contamination and the microbiological profile of fresh broiler chicken carcasses using the Nu-Tech Evisceration System or the conventional Streamlined Inspection System, *Poultry Science*, 76, 780–784.

Sams A R (2001), First processing: Slaughter through chilling, in *Poultry Meat Processing*, ed A Sams, CRC Press, Boca Raton, FL, 19–34.

Smith D P, Northcutt J K, Cason J A, Hinton A, Jr, Buhr R J, and Ingram K D (2007), Effect of external or internal fecal contamination on numbers of bacteria on prechilled broiler carcasses, *Poultry Science*, 86, 1241–1244.

U.S. Department of Agriculture Food, Safety Inspection Service (2005), Verification of procedures for controlling fecal material, ingesta, and milk in slaughter operations, FSIS Directive 6420, Washington, DC.

Chapter 18

Reducing *Salmonella* on the Processing Line Using Carcass Sprays

18.1 Introduction

Reducing pathogens on the processing line involves two separate processes. One is the reduction of cross contamination from carcass to equipment and vice versa. The other is reduction of bacteria on carcasses using sprays or dips.

Arnold and Silvers (2000) reported that mechanical equipment has vastly increased the number of carcasses processed by a single plant each day. Movement of the industry toward automation has resulted in the presentation of new surface areas for carcasses to contact repeatedly and thus new opportunities for bacterial attachment and cross contamination (McEldowney and Fletcher, 1988).

The purpose of equipment sprayers is to disinfect the part of the processing equipment that comes into contact with each carcass to prevent pathogens from coming off one carcass and being transferred to another carcass, resulting in cross contamination. Moreover, the nozzles and bars should be positioned so that they spray the part of the equipment that touches the chicken carcass and not other parts. Used properly, these systems can help to reduce cross contamination. Generally, high-pressure spray nozzles are used in these applications, and by far the most commonly used chemical is chlorine because of its cost. Other chemicals that are being used for these applications include chlorine dioxide, peracetic acid, and Zentox (monochloramine) or TOMCO water (acidified hypochlorous acid). While chlorine is an excellent sanitizer, there are some general principles that must be taken into account when using it. The incoming water pH should be less than 6.5 after addition of the sodium hypochlorite bleach for the chlorine to do its job (see Chapter 22). This may be accomplished by the addition of citric acid or carbon dioxide to the water.

18.2 Effect of Organic Material

The main inhibitor to chlorine being effective as an equipment disinfectant is the buildup of fat or other organic material on the equipment, such that even though chlorinated water is being sprayed onto the built-up material, it is not able to penetrate the material and kill the bacteria underneath. These bacteria may then be liberated when the next carcass comes by and are then transferred to the subsequent carcass. This is a common occurrence on carcass brushes intended to remove feces after evisceration. Often, even though chlorinated sprays are used, these systems end up causing an increase in *Salmonella* prevalence when biomapping is done on carcasses before and after the brush or fecal finger system. The cropper can be especially troublesome in this regard in that the ingesta in the crops of birds often contain pathogens such as *Salmonella* and *Campylobacter*.

18.3 Effect of pH

The pH of the equipment rinse waters should be maintained below 6.5 and checked regularly to ensure that the chlorine is in the appropriate form for optimal activity against bacteria. The spray pressure should also be checked and any clogged nozzles should be replaced. Figure 18.1 shows a spray bar in which the nozzles are partially plugged (reducing pressure), and some of the nozzles are not working at all.

In some cases, a maintenance person comes along and changes the nozzles or turns down the pressure without understanding the microbiological effect on the carcass. This can have a dramatic negative impact on the ability of the sprayer to be effective.

General suggestions for all washers and rinsers include

1. Maintain proper nozzle pressure
2. Maintain proper water pH
3. Maintain proper chlorine or other chemical level
4. Maintain proper water distribution on the carcass

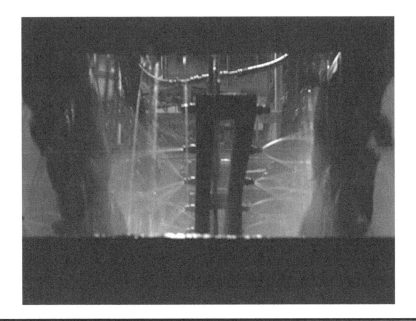

Figure 18.1   An improperly operating spray bar with nozzles plugged.

18.4 Effect of Other Chemicals

Yang et al. (1998) conducted a study in which antimicrobial sprays were applied using a modified inside-outside bird washer (IOBW) to reduce *Salmonella typhimurium* and aerobic plate counts (APCs) on prechilled chicken carcasses in a poultry-processing pilot plant. Four chemicals—trisodium phosphate (TSP, 10%), lactic acid (LAC, 2%), cetylpyridinium chloride (CPC, 0.5%), and sodium bisulfate (SBS, 5%)—were selected to be tested as antimicrobial agents. Each chicken carcass was inoculated by spraying the inside and outside with *Salmonella typhimurium* at 10^5 colony-forming units (CFU) per carcass. The inoculated carcasses then were passed through the bird washer and sprayed with each chemical at 35°C for 17 seconds. Yang et al. (1998) found that all of the chemical treatments reduced *Salmonella* on the chicken carcasses by approximately 2 $\log_{10}$ CFU per carcass. Total aerobes on the chicken carcasses, however, were reduced by 2.16, 1.66, 1.03, and 0.74 $\log_{10}$ CFU per carcass after spraying with 0.5% CPC, 5% SBS, 2% LAC, or 10% TSP, respectively. The authors concluded that the most effective antimicrobial spray treatment for reducing both *Salmonella* and total aerobes on prechilled chicken carcasses was 0.5% CPC.

In most cases, companies generally use chlorine in IOBW systems; however, a comprehensive research study conducted by Abu-Ruwaida et al. (1994) showed that spray washing after evisceration had no effect on levels of bacteria. Likewise, a study by Northcutt et al. (2005) of the Agricultural Research Service (ARS) of the U.S. Department of Agriculture (USDA) was conducted to investigate the effects of spray washing broiler carcasses with acidified electrolyzed oxidizing (EO) water or sodium hypochlorite (NaOCl) solutions for 5, 10, or 15 seconds. Commercial broiler carcasses were contaminated with 0.1 g of broiler cecal contents inoculated with 10^5 cells of *Campylobacter* and 10^5 cells of nalidixic acid-resistant *Salmonella*. Northcutt et al. (2005) concluded that adding chlorine (50 ppm) to the IOBW has absolutely no impact on APCs, *E. coli* counts, *Salmonella* prevalence, or *Campylobacter* counts on carcasses. These data correlated well with numerous biomapping studies that showed no difference in the microbiological quality of carcasses entering versus those exiting the IOBW systems using 50 ppm chlorine. These results are generally surprising when presented to industry personnel; however, this may be explained by the high level of organic material on the carcass at that location in the process. Chlorine may not be able to penetrate the organic material and interact with the *Salmonella* attached to the surface of the carcass.

References

Abu-Ruwaida A S, Sawaya W N, Dashti B H, Murad M, and Al-Othman H A (1994), Microbiological quality of broilers during processing in a modern commercial slaughterhouse in Kuwait, *Journal of Food Protection*, 57, 887–892.

Arnold J W, and Silvers S (2000), Comparison of poultry processing equipment surfaces for susceptibility to bacterial attachment and biofilm formation, *Poultry Science*, 79, 1215–1221.

McEldowney S, and Fletcher M (1988), Bacterial desorption from food container and food processing surfaces, *Microbial Ecology*, 15, 229–237.

Northcutt J K, Smith D P, Musgrove M T, Ingram K D, and Hinton A, Jr (2005), Microbiological impact of spray washing broiler carcasses using different chlorine concentrations and water temperatures, *Poultry Science*, 84, 1648–1652.

Yang Z, Li Y, and Slavik M (1998), Use of an antimicrobial spray applied with an inside-outside bird washer to reduce bacterial contamination on pre-chilled chicken carcasses, *Journal of Food Protection*, 61, 829–832.

Chapter 19

Effect of Online Reprocessing on *Salmonella* on Carcasses

19.1 Introduction and History of Online Reprocessing

According to an excellent review of the history of online reprocessing (OLR) by Bailey (2004) of the Agricultural Research Service (ARS) of the U.S. Department of Agriculture, approximately 0.5–1% of processed broilers require reprocessing. This percentage equates to 45 to 90 million carcasses per year. This figure is variable based on the plant and the time of year. In some instances, some plants have had to reprocess up to 12% of their carcasses because their evisceration equipment was not adjusted properly, and intestines were being torn during evisceration. Prior to 1989, the USDA-FSIS (Food Safety Inspection Service) would allow a fecally contaminated carcass to be inspection passed if the part that had feces on it was trimmed from the carcass (Russell, 2007). However, if the fecal contamination occurred on an internal surface, the carcass could not be trimmed and hence would not pass inspection (Bailey, 2004). In 1989, the *Code of Federal Regulations* was amended such that, under the supervision of a USDA inspector, reprocessing treatments, including trimming, vacuuming, washing, or a combination, were allowed. If internal contamination was present or treatments other than trimming were to be used, then the entire carcass must be washed with water containing 20 ppm chlorine (Bailey, 2004).

19.2 Effect of Off-line, Manual Reprocessing

After the new regulation went into effect, Blankenship et al. reevaluated off-line reprocessing in 1993 in five different poultry-processing facilities and found that inspection-passed carcasses were microbiologically indistinguishable from fecally contaminated and manually reprocessed (off-line reprocessed) carcasses (Figures 19.1 and 19.2).

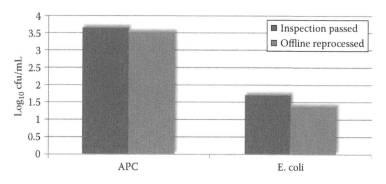

Figure 19.1 Log_{10} **colony-forming units (CFU) per milliliter of APC and** *E. coli* **on chicken carcasses that were inspection passed versus those that were contaminated with feces and manually reprocessed using chlorinated water. (From Blankenship L C, Bailey J S, Cox N A, Stern N J, Brewer R, and Williams O, 1993, Two-step mucosal competitive exclusion flora treatment to diminish salmonellae in commercial broiler chickens,** *Poultry Science,* **72, 1667–1672.)**

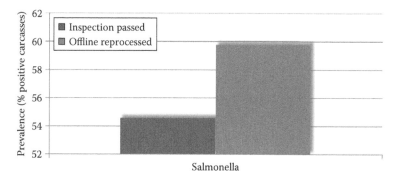

Figure 19.2 **Prevalence of** *Salmonella* **on chicken carcasses that were inspection passed versus those that were contaminated with feces and manually reprocessed using chlorinated water. These data are averages over five poultry plants and were not calculated to be significantly different** ($p < 0.05$). **(From Blankenship L C, Bailey J S, Cox N A, Stern N J, Brewer R, and Williams O, 1993, Two-step mucosal competitive exclusion flora treatment to diminish salmonellae in commercial broiler chickens,** *Poultry Science,* **72, 1667–1672.)**

These data indicate that poultry processors were able to take fecally contaminated carcasses and manually wash and sanitize them such that they were microbiologically equivalent to carcasses that were inspection passed with no fecal contamination (Russell, 2007). In the same year, Waldroup et al. (1993) conducted a study on manually reprocessed broilers and found that, although some variability was noted between plants, *Salmonella* prevalence and numbers were not different for inspection-passed broilers when compared to fecally contaminated and manually reprocessed broilers. Waldroup et al. also reported that *Campylobacter* numbers on manually reprocessed broilers were the same as or, in some cases, lower than those on inspection-passed broilers.

19.3 Effect of Online Reprocessing

In 1997, Fletcher et al. conducted a landmark study that demonstrated that if broilers were processed using an OLR system instead of manually reprocessing the broilers, no differences could be observed in aerobic plate count (APC), *Salmonella*, or *Campylobacter* numbers. Fletcher et al. (1997) concluded that OLR of visually contaminated carcasses could greatly reduce the number of carcasses subjected to off-line reprocessing without negatively affecting bacterial counts and, specifically, *Salmonella* and *Campylobacter*. This helped to usher in the new era of OLR. When companies begin using OLR systems, although the OLR chemicals are often expensive to use, the companies justify the cost of the OLR system because they are not spending as much on labor to reprocess the chickens manually. Likewise, they do not have to deal with labor issues related to those individuals who do the manual reprocessing, such as absenteeism, repetitive movement injuries, and the like.

19.4 Online Reprocessing Chemicals

Since that time, some of the chemicals that have been approved for OLR applications include trisodium phosphate (TSP); acidified sodium chlorite (ASC; Ecolab Sanova®); peracetic acid (PAA; FMC 323) and combinations of PAA and octanoic acid (Inspexx 100), chlorine dioxide (multiple suppliers); acidified chlorine (TOMCO); cetylpyridinium chloride (CPC; SafeFoods-Cecure); mixtures of acids (AFTEC 3000, FreshFx); hydantoinated bromine (Bromatize); and acidified calcium sulfate (Safe2O). All of these chemicals are currently used in the poultry industry in OLR applications. In most applications, these chemicals are applied using sprayers (in some cases, sophisticated spray systems, such as the Chad cabinet), or deluge systems (Figure 19.3), which literally flush the carcass with the solution. In some cases, the total OLR application procedure is linked with the inside-outside bird washer (IOBW), as is the case with the TOMCO application.

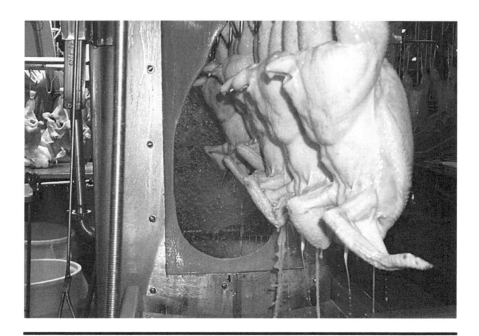

Figure 19.3 Deluge OLR system used to deliver TSP to carcasses.

Kemp and others in 2001 reported that ASC (Sanova, now sold by Ecolab) performed exceedingly well. The researchers reported that fecally contaminated carcasses that were treated with Sanova had significantly lower microbial levels than those that were fecally contaminated and manually reprocessed. In fact, *Escherichia coli* and *Campylobacter* were reduced by 1.78 and 1.75 $\log_{10}$ more when online reprocessed as compared with manual reprocessing, and *Salmonella* prevalence was 21.6% less for the online-processed carcasses. These data suggest that OLR is far superior to off-line manual reprocessing. Moreover, the online-reprocessed chickens were microbiologically superior to inspection-passed carcasses. This is a huge incentive for the USDA-FSIS to encourage the use of OLR.

19.5 Reasons Why Online Reprocessing Is Superior to Off-Line, Manual Reprocessing

What factors may be responsible for enabling OLR to be so superior to off-line manual reprocessing? With off-line reprocessing, only chlorinated water is used. Chlorinated water is insufficient to penetrate the biofilms on the surface of the chicken, and in general, the "higher-tech" chemistries are superior to chlorine with regard to bacterial killing power and their ability to remain active on the surface of the carcass, whereas chlorine becomes inactive as soon as it contacts the skin of the chicken. Another important factor is that the USDA-FSIS requires that salvaged carcasses remain in the salvage area until they can be inspected to allow them to be released to the chiller. Usually, carcasses are stored in a warm area, and the time required for inspectors to come to the salvage area and inspect the carcasses may be between 30 and 45 minutes, allowing bacteria on the surface of the chicken to multiply during that time. The reasons listed are often used for dissuading a processor from using manual reprocessing, along with the labor expense associated with the employees required to manually reprocess the chickens.

It can be concluded that these systems have been scientifically demonstrated to be equal to or significantly better than off-line reprocessing because these were the criteria used for their approval by the USDA-FSIS. The USDA-FSIS requirement for a company to receive a "letter of no objection" for use of its product as an OLR agent is complex and expensive to achieve. In some cases, hundreds of thousands of dollars have been spent achieving such a letter from the USDA-FSIS. This begs the question: How could the USDA-FSIS justify allowing companies to spend this much money achieving a letter of no objection and then pull the rug out from under the company and disallow the practice of allowing OLR altogether?

19.6 USDA-FSIS View of Online Reprocessing

Even though OLR systems have been shown to work well, on April 3, 2007, the USDA-FSIS issued the following letter to the companies that sell and distribute OLR chemistries (Russell, 2007):

> Dear On-Line Reprocessing Provider:
>
> Pending Agency amendment of §381.91 (b) (1) [regulation on off-line reprocessing], the use of your On-Line Reprocessing (OLR) system continues pursuant to a waiver that Food Safety and Inspection Service (FSIS) had issued under §381.3 (b).
>
> Since you are operating under a waiver of a regulation, the FSIS New Technology Staff (NTS) must ensure the on-going effectiveness of your OLR system including evidence that it continues to have the effect of improving the microbial quality of reprocessed birds through monitoring data at a regular frequency.
>
> NTS has not received recent OLR monitoring data from you. Therefore, we request you to submit monitoring data by May 5, 2007.
>
> In particular, the Agency requests microbial results comparing A) rinses of poultry carcasses that were free of visible contamination, and B) rinses of carcasses that had visible fecal contamination after going through standard OLR treatment but before they enter the chilling system.

This letter means that the bar has been raised. In the future, for each company to achieve approval for the use of their chemistry in an OLR capacity, they must prove that carcasses that were fecally contaminated and went through the OLR system are microbiologically equivalent to or better than carcasses that were inspection passed that went through the OLR system. This is completely different from requiring that OLR systems be able to reduce bacteria on fecally contaminated carcasses to a level that is equal to or better than a carcass that was inspection passed. Figures 19.4 and 19.5 clearly demonstrate the problem with this model compared to the former standard. Figure 19.5 shows the new standard that would be required if implemented.

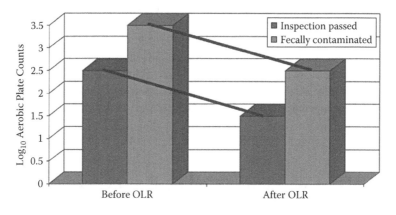

Figure 19.4 **Log_{10} APCs must be reduced by 1 when traversing the OLR for both inspection-passed and fecally contaminated carcasses.**

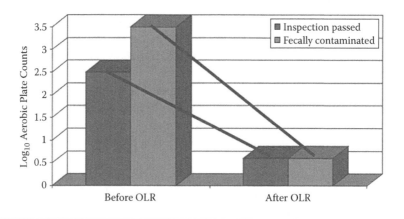

Figure 19.5 **The new requirement is that log_{10} APCs must be microbiologically equal when traversing the OLR for both inspection-passed and fecally contaminated carcasses.**

In Figure 19.4, even though the number of bacteria on the carcasses that were fecally contaminated that exited the OLR system was equivalent to the number of bacteria on carcasses that were inspection passed before OLR, most chemistries are not able to reduce the fecally contaminated carcass bacterial numbers to a greater degree, so that the final OLR bacterial counts for both groups of carcasses (fecal and inspection passed) are equal as in Figure 19.5. This is because of a common tenet in microbiology: Junk in equals junk out. If the OLR is expected to reduce bacterial levels by 1 $\log_{10}$ and the OLR is working as expected in Figure 19.4, then how can carcasses that start at $\log_{10}$ 3.5 be reduced 2 logs to equal the inspection-passed carcasses? It is likely that very few of the companies that supply OLR chemistries to the poultry industry will be able to achieve the new requirement. In reviewing data from over 2,700 carcasses that were processed using a variety of OLR systems in numerous poultry plants, I drew a similar conclusion. Therefore, the new regulations by the USDA-FSIS may effectively shut down the use of OLR systems nationwide. Also, the new rule completely ignores the original purpose for OLR systems: to make fecally contaminated carcasses microbiologically equivalent to those that are inspection passed. If the USDA-FSIS continues along this path, OLR systems will likely become a thing of the past.

This new requirement creates numerous questions, such as

- What happens to all of the fecally contaminated carcasses?
- Will the companies have to begin trimming or manually reprocessing carcasses again?
- What is the cost of this immense reversal back to the older methods?
  - What about the companies that supply OLR chemistries that have millions of dollars invested in these technologies in testing and infrastructure that are no longer able to meet the new discriminatory standard?
  - What about the poultry companies that have spent millions of dollars in capital costs to install OLR systems that are now extinct?

A scientific reason should be given by the USDA-FSIS for making this immense change to provide reasoning to the poultry industry and the OLR suppliers regarding why this change is necessary to protect the public. It is my suspicion that this regulation is a response to consumer groups in Washington, D.C., complaining about "animal feces on food," and that this is unacceptable. These groups are unwilling to examine scientific data because their ultimate agenda is to eliminate animal food production completely. It is extremely important that the governmental agencies tasked with the safety of food in the United States use scientific data to make decisions as opposed to responding to people who have no interest in food safety. This change may make poultry less safe, as indicated by the reams of data and scientific studies that indicate that manually reprocessed carcasses are more contaminated than carcasses processed using OLR chemistries.

19.7 Purpose of Online Reprocessing

The purpose of OLR systems is not to reduce *Salmonella*. The USDA-FSIS views the OLR as a process intended to make carcasses that would otherwise have to be reprocessed by hand because they have fecal material or ingesta on them microbiologically equivalent to those carcasses that do not have any fecal material or ingesta on them. This is, in fact, the type of research protocol that companies must run to achieve approval for their chemical as an OLR agent (i.e., they must compare the microbiological quality of fecally contaminated to uncontaminated carcasses). That having been said, most processors expect to achieve at least a 1 $\log_{10}$ reduction in bacterial levels on carcasses as they traverse the OLR system.

19.8 Online Reprocessing Chemicals

The U.S. Poultry and Egg Association conducted a survey of the poultry industry in February 2006, and those data were presented by Rice (2006). The following are chemicals that the poultry industry (94 plants responded to the survey out of approximately 247 in the United States) is using for OLR purposes and the percentage of companies that use that particular chemistry: ASC (Sanova, 33%); TSP (Rhodia, 24%); chlorine dioxide (numerous companies, 15%); hypochlorous acid (Zentox and TOMCO, 9%); organic acids (6%); PAA (FMC 323 or Parasafe and Inspexx 100, 5%); CPC (Safefoods Cecure®, 3%); SynerX® (citric acid and HCL, 1%); bromine (Bromitize™, 1%); sodium metasilicate (AvguardXP®, 1%); and electrolyzed oxidative (EO) water (EAU, 1%). Chemicals not mentioned in the survey include Zentox monochloramine and SteriFx (FreshFx), which was included with organic acids but contains mostly inorganic acid.

19.8.1 Acidified Sodium Chlorite

Acidified sodium chlorite (ASC) is approved as a poultry spray or dip at 500 to 1,200 ppm singly or in combination with other generally recognized as safe (GRAS) acids to achieve a pH of 2.3 to 2.9 as an automated reprocessing method. In chiller water, sodium chlorite is limited to 50 to 150 ppm singly or in combination with other GRAS acids to achieve a pH of 2.8 to 3.2 (Russell, 2007).

In 2001, Kemp et al. determined the effectiveness of the combined use of an IOBW for the removal of visible contamination and an online ASC spray system in reducing microbial levels on contaminated poultry carcasses. Carcasses were sampled for *Escherichia coli*, *Salmonella*, and *Campylobacter* at five stations along the processing lines in a series of five commercial plant studies to compare the efficacy of the OLR to that of off-line processing. The microbiological quality of fecally contaminated carcasses was found to be significantly better following the OLR treatment (Kemp et al. 2001; Figure 19.6).

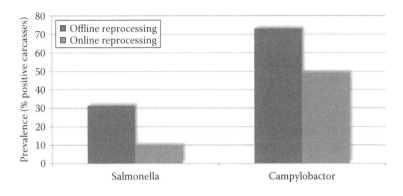

Figure 19.6 Prevalence of *Salmonella* and *Campylobacter* on chicken carcasses using off-line reprocessing with chlorine or online reprocessing using acidified sodium chlorite. (Sanova. From Kemp G K, Aldrich M L, Guerra M L, and Schneider K R, 2001, Continuous online processing of fecal- and ingesta-contaminated poultry carcasses using an acidified sodium chlorite antimicrobial intervention, *Journal of Food Protection*, 64, 807–812.)

Sexton et al. (2007) evaluated the effectiveness of ASC on *Salmonella* and *Campylobacter* on chicken carcasses after they exited the screw chiller in Adelaide, Australia. For untreated carcasses, the average $\log_{10}$ APC was $2.78/cm^2$ compared with $1.23/cm^2$ for treated carcasses. Prevalence of *E. coli, Salmonella,* and *Campylobacter* was 100%, 90%, and 100% respectively, on untreated carcasses and 13%, 10%, and 23%, respectively, on treated carcasses (Sexton et al., 2007). The authors reported that the significant reductions in bacterial numbers and prevalence indicate that ASC for use in OLR systems is effective.

Del Rio et al. (2007) tested the effects of dipping carcasses (15 minutes) in potable water or in a solution (w/v) of 1,200 ppm ASC on *Salmonella* Enteritidis. ASC reduced microbial populations ($p < 0.001$) as compared with the control (untreated) samples. Actual APC reductions ranged from 0.33 to 3.15 log colony-forming units (CFU)/gram using ASC (Del Rio et al., 2007).

19.8.2 Trisodium Phosphate

Use of TSP was originally encouraged by the USDA-FSIS as an approved method for OLR (Russell, 2007). TSP is costly to use because of the high concentration (10%) used on carcasses. There are negative aspects to using TSP in poultry-processing plants that should be considered. Residual TSP on carcasses causes the chiller water pH to increase dramatically. In plants where TSP is used, the chiller water will generally be in the pH range of 9.7 to 10.5. This is extremely high and prevents chlorine from being converted to its effective form, hypochlorous acid (see Chapter 22). Hypochlorous acid forms most effectively when water is at a pH below

6.5. Thus, plants using TSP are wasting their bleach. This is not a desired situation because chlorine is very effective against *Salmonella.*

If a poultry company is having trouble with high *Salmonella* prevalence and has an operating TSP system in place, it must make major adjustments to reduce *Salmonella* prevalence. CO_2 gas systems have been added to the aeration systems of chillers as a means of reducing the pH of the water so that when chlorine is added, it will form hypochlorous acid. Discharging TSP in areas of the country that have strict phosphorous discharge limitations may be a problem as well. One beneficial effect of using TSP is that companies report that a 1% yield increase may be achieved due to increased water-holding capacity.

Coppen et al. (1998) studied the efficacy of the AvGard TSP immersion carcass wash process in industrial trials against *Salmonella*. Dramatic reductions in *Salmonella* prevalence were observed using a whole-carcass rinse method. The authors reported that, in a water-chilled broiler plant, an average control prevalence for *Salmonella* of 74.0% was reduced to 9.4% after AvGard treatment (Coppen et al., 1998). It is important to note that great care must be taken when conducting these types of studies because, when using carcass rinses, TSP is rinsed off the carcass and will artificially kill *Salmonella* in the rinsate. Without careful neutralization of the effect of TSP in the rinsate, significant artificial reductions will be observed.

19.8.3 Chlorine Dioxide

Chlorine dioxide has had a rocky past within the poultry industry. Early attempts to introduce this chemical were unsuccessful because of the inability to control the levels of ClO_2 during use. Gassing off occurred frequently, and employees complained. ClO_2 is an extremely effective sanitizer. The companies that are most successful with this chemical produce the chemical on site and control it carefully. Adequate ventilation is necessary to ensure worker safety. In an in-plant trial conducted recently, we found no observable reduction in APC or *E. coli* on carcasses as they traversed an OLR using ClO_2. This may be related to the relatively short contact time used in this plant and should not be used to evaluate all ClO_2 used in all OLR applications.

19.8.4 Hypochlorous Acid

Hypochlorous acid is formed when bleach is dissolved in water. The lower the pH (in particular below 6.5) is, the higher the concentration of hypochlorous acid will be. The success of this technology varies greatly as well. Companies have seen excellent-to-no positive results depending greatly on the organic loading of the carcasses entering the system. In systems that adequately clean the carcasses prior to introduction to the OLR system, the bacteria may be greatly reduced using this approach. However, carcasses that have high organic loading in plants that use few carcass washes and little water may not achieve success using HOCl.

19.8.5 Organic or Inorganic Acids and Mixtures of Acids

Acids definitely kill bacteria; however, they must be closely monitored to ensure that they contact the skin of the carcasses for an appropriate period of time and that they do not create product defects. Acids need more time to kill bacteria than oxidant-based chemicals, and if the chiller immediately (usually within 2 minutes) follows the OLR system, it can be difficult to achieve good results. Often, bacteria become acid stressed when a carcass is treated with an acid. These organisms become hard to recover when doing efficacy studies. This does not mean that the bacteria were killed and will not be discovered by the USDA. Thus, when using acids, it is important to make sure that adequate neutralization and recovery steps are used during microbiological analysis or inaccurate results will be obtained.

19.8.6 Peracetic Acid

Peracetic acid (PAA) is a mixture of an organic acid (acetic acid) and an oxidant (hydrogen peroxide). Therefore, this chemical kills bacteria in two separate ways. Most research indicates that a 1 $\log_{10}$ reduction may be achieved using approximately 200 ppm PAA in the OLR system. A precaution when using PAA is that it may react with blood vessels, producing a slightly gray color on the skin of the carcasses in areas that are highly vascularized, such as the neck. PAA is an extremely effective antimicrobial agent.

19.8.7 Cetylpyridinium Chloride

Cetylpyridinium chloride (CPC) is an effective OLR technology. We have conducted studies that have demonstrated an 83% reduction in *Salmonella* on carcasses traversing this system. Unfortunately, in another plant, we achieved only a 10% reduction using this method, even though the exact same spray cabinet, pressure, and concentration of chemical were used. The reason why the chemical did not work as well in the second situation is that the bacteria on the surface of the carcasses were firmly encased in biofilms, and the CPC was not able to access the organisms to eliminate them. If the company takes steps to ensure that the biofilm on the surface of the chicken is disrupted prior to treatment with CPC, then the treatment should be very effective.

19.8.8 Electrolyzed Oxidative Water

The machine used to generate EO water is expensive; however, the cost of the raw materials is very low (salt and water). Thus, the total cost is reasonable. Studies have shown that it is very effective and can achieve a 1 $\log_{10}$ reduction in bacteria on carcasses while being completely safe to use. EO water is acidified (pH 1.9 to 2.4) oxidative water that contains some hypochlorous acid (50 ppm) and other antimicrobial ions. It is generated on site, stored, and used as generated. It is not diluted. This material is excellent for postchill dip applications as well.

19.8.9 Monochloramine

Monochloramine (Zentox) has many of the advantages of chlorine without the negative aspects. Monochloramine is used in a similar fashion as chlorine at 50 ppm. It is generated by mixing bleach and ammonia under controlled conditions. It kills bacteria but is resistant to deactivation by organic material. Thus, it is more stable under high organic loads. Axtell et al. (2006) demonstrated that no carcinogenic compounds were formed when monochloramine was used to chill poultry carcasses. This is a major concern in Europe with regard to the use of chlorine. In fact, monochloramine is used as a drinking water disinfectant in Europe.

Unfortunately, there are still no magic bullets for use with OLR systems. The short contact time and methods of application (generally a spray) make it difficult for chemicals to eliminate *Salmonella* during this step. We have observed reductions in *Salmonella* prevalence of 0–90% using various chemistries. It is also possible to see this type of variation from plant to plant using an individual chemical. This is because plants and incoming loads vary tremendously from plant to plant. Thus, it is important to select the appropriate type of sanitizer based on an individual plant setup.

References

Axtell S P, Russell S M, and Berman E (2006), Effect of immersion chilling of broiler chicken carcasses in monochloramine on lipid oxidation and halogenated residual compound formation, *Journal of Food Protection*, 69, 907–911.

Bailey J S (2004), Reprocessing of fecal contaminated carcasses and the use of antimicrobials. http://www.fsis.usda.gov/PDF/Slides _022306_S Bailey2.pdf.

Blankenship L C, Bailey J S, Cox N A, Stern N J, Brewer R, and Williams O (1993), Two-step mucosal competitive exclusion flora treatment to diminish salmonellae in commercial broiler chickens, *Poultry Science*, 72, 1667–1672.

Coppen P, Fenner S, and Salvat G (1998), Antimicrobial efficacy of AvGard carcass wash under industrial processing conditions, *British Poultry Science*, 39, 229–234.

Del Rio E, Muriente R, Prieto M, Alonso-Calleja C, and Capita R (2007), Effectiveness of trisodium phosphate, acidified sodium chlorite, citric acid, and peroxyacids against pathogenic bacteria on poultry during refrigerated storage, *Journal of Food Protection*, 70, 2063–2071.

Fletcher D L, Craig E W, and Arnold J W (1997), An evaluation of on-line reprocessing on visual contamination and microbiological quality of broilers, *Journal of Applied Poultry Research*, 6, 436–442.

Kemp G K, Aldrich M L, Guerra M L, and Schneider K R (2001), Continuous online processing of fecal- and ingesta-contaminated poultry carcasses using an acidified sodium chlorite antimicrobial intervention, *Journal of Food Protection*, 64, 807–812.

Rice J (2006), *Salmonella* interventions in the broiler industry. http://www.fsis.usda.gov/PDF/Slides_022406_EKrushinskie.pdf.

Russell S M (2007), A bar too high? *PoultryUSA Magazine*, July, 32–34.

Sexton M, Raven G, Holds G, Pointon A, Kiermeier A, and Sumner J (2007), Effect of acidified sodium chlorite treatment on chicken carcasses processed in South Australia, *International Journal of Food Microbiology*, 30, 252–255.

Waldroup A L, Rathgeber B M, Heirholzer R E, Smoot L, Martin L M, Bilgili S F, Fletcher D L, Chen T C, and Wabeck C J (1993), Effects of reprocessing on microbiological quality of commercial pre-chill broiler carcasses, *Journal of Applied Poultry Research*, 2, 111–116.

Chapter 20

Effect of Immersion Chilling on *Salmonella*

20.1 Introduction

There has been a great debate for many years over the benefits and disadvantages of immersion chilling versus air chilling. Most plants in the United States use immersion chilling, whereas most plants in Europe use air chilling.

20.2 Effect of Immersion Chilling on *Salmonella* Levels

A major criticism of immersion chilling is that, in the past, companies did not use proper management techniques and chemical intervention to prevent cross contamination with *Salmonella*. Smith et al. (2005) studied the effect of prechill fecal contamination on numbers of bacteria on immersion-chilled carcasses. Cecal contents (0.1 g inoculated with *Campylobacter* and nalidixic acid-resistant *Salmonella*) were applied to each of eight carcass halves in one group (direct contamination) that were placed into one paddle chiller (contaminated), whereas the other paired halves were placed into another chiller (control). From the second group of eight split birds, one of each paired half was placed in the contaminated chiller (to determine cross contamination), and the other half was placed in the control chiller. There were no significant statistical differences from direct contamination for coliforms and *Escherichia coli*, although *Campylobacter* numbers significantly increased from control values because of direct contamination, and the prevalence increased from 79% to 100%. There was no significant effect of cross contamination on coliform or *E. coli* numbers. *Campylobacter* levels were significantly higher after exposure to cross contamination, and the incidence of this bacterium increased from 75% to 100%. *Salmonella*-positive halves increased from 0% to 42% postchill because of direct contamination and from 0% to 25% as a result of cross contamination after chilling. Smith et al. (2005) concluded that *Campylobacter* numbers, *Campylobacter* incidence, and *Salmonella* incidence increased because of both direct contamination and cross contamination in the chiller. Postchill *E. coli* numbers did not indicate which carcass halves were contaminated with feces before chilling.

This study served to demonstrate that the criticism associated with immersion chillers concerning potential for cross contamination is very real. It is important to note that when immersion chillers are improperly controlled with regard to countercurrent flow, adequate freshwater input, control of chlorine level to ensure a free available chlorine content of 1 to 5 ppm, and pH adjustment to below 6.5, cross contamination does not occur.

Reference

Smith D P, Northcutt J K, and Musgrove M T (2005), Microbiology of contaminated or visibly clean broiler carcasses processed with an inside-outside bird washer, *International Journal of Poultry Science*, 4, 955–958.

Chapter 21

Reducing *Salmonella* during Immersion Chilling

21.1 Introduction

More bacterial reduction (both numbers and prevalence) can be accomplished in a properly balanced chiller than anywhere else in the processing plant (Russell, 2007). Most studies demonstrated that the chiller can significantly reduce *Salmonella* prevalence (Izat et al., 1989) if operating properly. As with the scalder, the pH, flow rate, flow direction, chlorine concentration, and concentration of organic material (digesta, fat, blood) are crucial for the chlorine in the chiller to adequately prevent cross contamination from carcass to carcass and to lower *Salmonella* numbers on carcasses (Russell, 2007). The pH should be below 6.5, the flow rate should be high (at least 2 L per carcass), and the flow direction should be countercurrent. The most effective methods for controlling the pH of chill water include addition of carbon dioxide gas (90% of the industry uses this method according to the U.S. Poultry and Egg Association survey presented by Rice, 2006) to the tubes normally used for air agitation, and the addition of citric acid (10% of the industry uses this according to the survey) or other acidifiers such as sodium acid sulfate to the water.

21.2 Controlling Organic Loading in Immersion Chillers

The organic material in an immersion chiller is generally determined by the following factors: use of prescald bird brushes to remove fecal material on birds prior to entering the scalder; bleed-out efficiency (determined by stunning); temperature of the scalder (higher temperatures melt subskin fat, which is released during chilling); flow rate in the scalder (amount of fresh makeup water); flow direction in the scalder (should be countercurrent); number of carcass sprays along the evisceration line; flow rate in the chiller (amount of fresh makeup water); flow direction in the chiller (should be countercurrent); and the number of chill tanks (more tanks are optimal). Excessive organic material (blood, ingesta, digesta, fat, protein) in the chiller will result in less chlorine being available to kill bacteria as it will be bound up and rendered useless by the organic material.

21.3 Effect of Organic Material on Free Chlorine in Chillers

Experiments conducted at the Western Regional Research Center, Agricultural Research Service, of the U.S. Department of Agriculture (USDA) concluded that a free chlorine residual could not be established in a commercial poultry chiller even by adding up to 400 ppm of free chlorine (Tsai et al., 1992). When chlorine reacts with organic material, it cannot act as a disinfectant because the chlorine is not free to oxidize the bacteria. Therefore, to maximize chlorine use in poultry chillers, efforts should be made to reduce the amount of organic material in these systems. Prescald bird brushes, effective carcass rinse systems, proper bleed-out procedures, countercurrent scalders and chillers, and proper freshwater makeup in scalders and chillers all contribute to lowering organic loading of the chillers.

21.4 Countercurrent Chilling

Many of the chillers in the poultry industry operate like a bath instead of a counterflowing river. The water appears stagnant, and organic material builds up during the shift (Figure 21.1). No gradient from dirty to clean water can be observed. Also, fat builds up on the chiller paddles, top of the water, and sides of the chiller (Figures 21.2–21.4). This allows *Salmonella* to be encased in the fat, offering it protection from the sanitizers used in the chiller.

A properly operating chiller should have a visible gradient such that the water at the chiller exit is significantly cleaner than the water at the entrance. This is

Figure 21.1 Chiller water containing high levels of organic material in the form of fat, blood, and protein.

Figure 21.2 Chiller water containing high levels of fat due to overscalding.

Figure 21.3 Exit of a chiller with water containing high levels of fat due to overscalding.

Figure 21.4 Another example of the exit of a chiller with water containing high levels of fat due to overscalding.

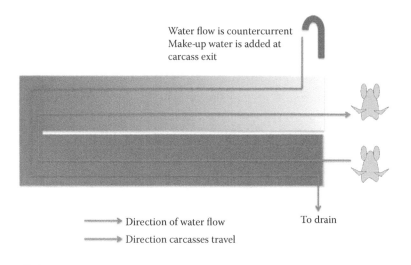

Figure 21.5   A diagram depicting an ideal chiller setup.

accomplished by adding all of the freshwater input and newly added chlorine directly to the exit end of the chiller as close to the exit paddles as possible. This suggestion is always met with the same statement by plant management and plant engineers: "We cannot chill the carcasses using this approach." However, this is untrue and is proved further in this chapter. This approach will result in a "clean space" near the exit end of the chiller. In this space, chlorine will be able to act against bacteria, similar to the way a postchill dip tank works. The ideal chiller setup is depicted in Figure 21.5.

The question is, can enough water be used to create this gradient in a large-scale plant? The answer is an emphatic, yes. A perfectly operating chiller may be seen in Figures 21.6–21.8 from two separate processing plants in Central and South America.

Even though all of the freshwater was being added to the exit end of the chiller and the redwater (water being removed from the chiller to be rechilled and added back to the chiller) was routed to enter the chiller at the center of the chill tank, the carcasses were able to be adequately chilled. These pictures prove that this approach may be used to achieve extremely effective chlorine disinfection and to chill carcasses adequately.

Many companies are now monitoring free available chlorine in the redwater. If the free available chlorine level is too low, chlorine is added to the cold redwater flowing back into the chiller until free available chlorine can be detected. This is usually done using an automatic feedback monitoring system and pump.

Figure 21.6 This is a chiller in El Salvador that is operating properly with extremely clean water containing very low organic loading.

Figure 21.7 This is a chiller in Chile that is operating perfectly with extremely low organic loading at the chiller exit.

Figure 21.8   This photo shows the enormous amount of fresh, cold, chlorinated water being added to the exit end of the chiller in Chile. This made the water in the chiller extremely clean and low in organic material.

21.5 Chemical Intervention on *Salmonella* in the Chiller

A number of chemicals have been approved for use in immersion poultry chillers; however, few are actually used. A survey in February 2006 by the U.S. Poultry and Egg Association and presented by Rice (2006) showed that the chemicals used in the United States for chiller applications and the percentage of plants that use them include hypochlorous acid (72%), peracetic acid (PAA; 18%), chlorine dioxide (8%), bromine (1%), and monochloramine (MON; 1%) (Figure 21.9).

Other chemicals not listed in the survey that have been used in the industry include sodium acid sulfate, citric acid, mixtures of organic and inorganic acids, and electrolyzed oxidative (EO) water. These chemicals all have advantages and disadvantages associated with their use, and some companies may find that one or the other is most appropriate for its specific plant environment. Overall, the chiller, if operated properly, can be the most significant intervention step for controlling *Salmonella* prevalence on broiler carcasses.

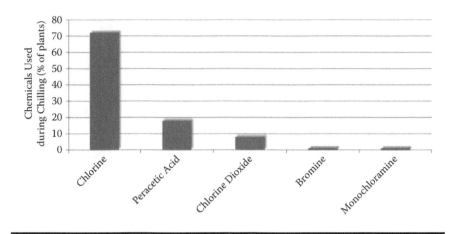

Figure 21.9   Chemicals used in commercial poultry chillers. (From Rice J, 2006, Salmonella interventions in the broiler industry. http://www.fsis.usda.gov/PDF/ Slides_022406_EKrushinskie.pdf.)

21.6 Cross Contamination during Immersion Chilling

In 1990, Lillard researched levels of aerobic bacteria, Enterobacteriaceae, and the prevalence of *Salmonella* pre- and postchill in a commercial processing plant. The level of aerobic bacteria and Enterobacteriaceae on broiler carcasses was reduced significantly by immersion chilling, but cross contamination still occurred. The authors found a significant increase in *Salmonella* prevalence on carcasses exiting the immersion chiller, indicating that this may be the point of most significant cross contamination in broiler-processing plants. This research demonstrated an essential problem with evaluating prevalence versus numbers of *Salmonella*. In this study, aerobic plate count (APC) and Enterobacteriaceae numbers were decreased significantly by immersion chilling, but the number of carcasses where a few *Salmonella* cells were transmitted from positive carcasses increased in number, even though many carcasses likely had significant reductions in *Salmonella* numbers. This begs the question: What is more important? Is it better to have 10% of carcasses with 10^5 *Salmonella* (infective dose) or 25% of carcasses with 5 cells of *Salmonella* (noninfective dose)? In the first case, this plant would be well within the *Salmonella* performance standard of the Food Safety Inspection Service (FSIS) of the USDA, but the consumer would be in danger of infection. In the latter case, the plant would not be in compliance with the USDA-FSIS standard, but none of the carcasses would be able to produce infection.

21.7 Effect of Sanitizers in the Chiller

21.7.1 Effect of Brifisol K

Rathgeber and Waldroup (1995) evaluated the bactericidal activity of Brifisol K (a commercial blend of sodium acid pyrophosphate and orthophosphoric acid) during immersion chilling of broiler carcasses. Brifisol K, at 1.5%, significantly reduced *Escherichia coli*, coliforms, and APCs on postchill broilers. Reductions in prevalence and levels of *Salmonella* were achieved.

21.7.2 Effect of Acetic Acid

Dickens and Whittemore (1995) studied the effects of extended chill times with and without 0.6% acetic acid (AA) and agitation on the microbiological quality of broiler carcasses. Carcasses were chilled for either 1, 2, or 3 hours using the following treatments: (1) paddle chiller without acid (C); (2) static ice slush with 0.6% AA (S); (3) static ice slush with air agitation and 0.6% AA (SA); and (4) a paddle-type chiller with 0.6% AA (P). The authors found that APCs were reduced by 0.34, 0.62, and 1.16 $\log_{10}$ most probable number (MPN)/mL for the S, SA, and P treatments, respectively, when compared with the controls. Enterobacteriaceae counts were reduced by 0.50, 0.71, and 1.4 $\log_{10}$ for the S, SA, and the P treatments, respectively. *Salmonella* prevalence, from inoculated carcasses, after 1 hour were 87% for the C carcasses, 80% for the S treatment, 53% for the SA treatment, and 6.7% for the P treatment. Thus, AA in a paddle chiller was able to reduce *Salmonella* by 81.3% more than tap water. It must be mentioned that AA, when used at levels that are effective, may impart a vinegar odor and flavor and a yellow-brown discoloration to the carcasses.

21.7.3 Effect of Sodium Hypochlorite, Acetic Acid, Trisodium Phosphate, and Sodium Metabisulfite

Tamblyn and Conner (1997) conducted experiments utilizing the skin attachment model (SAM) to determine the bactericidal activity of six potential carcass disinfectants (20, 400, and 800 ppm sodium hypochlorite [SH]; 5% AA; 8% trisodium phosphate [TSP]; and 1% sodium metabisulfite [SS]) during simulated immersion chilling for 60 minutes. All disinfectants except SS reduced numbers of freely suspended *Salmonella* Typhimurium by greater than or equal to 4.5 $\log_{10}$ colony-forming units (CFU)/mL. The SS did not reduce populations of salmonellae. SH at 20 ppm had little activity regardless of application, whereas higher levels were more effective, with bacterial populations reduced by 2.3 to 2.5 and 1.3 to 1.9 $\log_{10}$ CFU per skin, respectively. TSP was effective (reduction by 1.2 to 1.8 $\log_{10}$ CFU per skin) in all applications. AA was effective in the chiller application (2.5 log$_{10}$ reduction). The authors found that attachment of *S. typhimurium* to poultry skin apparently increased the ability of the bacteria to resist various disinfectants, and efficacy was influenced by extent of attachment of bacteria to skin and method of disinfectant application.

21.7.4 Effect of Electrolyzed Oxidative Water

In 1999, Yang et al. evaluated EO water for its antibacterial efficacy against *Salmonella typhimurium* on chicken carcasses during immersion chilling for 45 minutes. EO water with 50 ppm of oxidants in terms of free chlorine was used. During chilling, the 50-ppm chlorine EO water did not reduce *Salmonella* on carcasses but eliminated *Salmonella* in the chiller water (Yang et al., 1999).

21.7.5 Effect of Hydrogen Peroxide, Peracetic Acid, and Ozone

Vadhanasin et al. (2004) studied reduction in salmonellae achieved by Thai commercial exporters of frozen broiler chickens. *Salmonella* prevalence was 20.0% (6 of 30 samples) prior to washing and 22.7% (15 of 66) after chilling in tap water. Three corrective interventions were used during chilling: (1) 30 mg/L hydrogen peroxide, (2) 0.5% PAA, and (3) 125 mg/L ozone. The *Salmonella* prevalences after chilling were 16.0%, 5.0%, and 15.0% after hydrogen peroxide, PAA, and ozone treatments, respectively (Figure 21.10; Vadhanasin et al., 2004).

21.7.6 Effect of Monochloramine

Russell and Axtell (2005) conducted studies to compare the effect of SH versus MON on bacterial populations associated with broiler chicken carcasses. In Study 1, nominal populations (6.5 to 7.5 log CFU) of *Escherichia coli*, *Listeria monocytogenes*, *Pseudomonas fluorescens*, *Salmonella* serovars, *Shewanella putrefaciens*, and *Staphylococcus aureus* were exposed to sterilized chiller water (controls) or sterilized chiller water containing 50-ppm SH or MON. SH at 50 ppm eliminated all (6.5 to 7.5 log$_{10}$ CFU) viable *E. coli*, *L. monocytogenes*, and *Salmonella* serovars; 1.2 log$_{10}$ CFU of *P. fluorescens*; and 5.5 log$_{10}$ CFU of *S. putrefaciens*. MON eliminated all (6.5 to 7.5 log$_{10}$ CFU) viable *E. coli*, *L. monocytogenes*, *S. putrefaciens*, and *Salmonella* serovars and 4.2 log$_{10}$ CFU of *P. fluorescens*. In Study 2, chicken carcasses were inoculated with *P. fluorescens* or nalidixic acid-resistant *Salmonella* serovars or were temperature abused at 25°C for 2 hours to increase the populations of naturally occurring *E. coli*. The groups of *Salmonella* serovar-inoculated or temperature-abused *E. coli* carcasses were immersed separately in pilot-scale

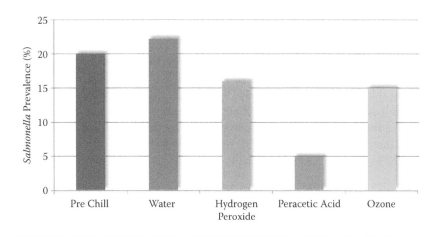

Figure 21.10 *Salmonella* **prevalence on broiler carcasses when disinfected during chilling using water, H₂O₂, peracetic acid, or ozone. (From Vadhanasin S, Bangtrakulnonth A, and Chidkrau T, 2004, Critical control points for monitoring salmonellae reduction in Thai commercial frozen broiler processing,** *Journal of Food Protection,* **67, 1480–1483.)**

poultry chillers and exposed to tap water (controls) or tap water containing 20-ppm SH or 20-ppm MON for 1 hour. The *P. fluorescens*-inoculated group was immersed in pilot-scale poultry chillers and exposed to tap water (controls) or tap water containing 50-ppm SH or 50-ppm MON for 1 hour. Carcasses exposed to the SH treatment had nominal increases (0.22 $\log_{10}$ CFU) in *E. coli* counts compared with controls, whereas exposure to MON resulted in a 0.89 $\log_{10}$ reduction. Similarly, average nalidixic acid-resistant *Salmonella* serovar counts increased nominally by 34% (41 to 55 CFU/mL) compared with controls on carcasses exposed to SH, whereas exposure to MON resulted in an average nominal decrease of 80% (41 to 8 CFU/mL). *Pseudomonas fluorescens* decreased by 0.64 $\log_{10}$ CFU on carcasses exposed to SH and decreased by 0.87 $\log_{10}$ CFU on carcasses exposed to MON. In Study 3, SH or MON was applied to the chiller in a commercial poultry-processing facility. *Escherichia coli* counts (for carcass halves emerging from both saddle and front-half chillers) and *Salmonella* prevalence were evaluated. Data from carcasses exposed to SH during an 84-day historical (Hist) and a 9-day prepilot (Pre) period were evaluated. Other carcasses were exposed to MON and tested during a 27-day period (Test). *Escherichia coli* counts for samples collected from the saddle chiller were 25.7, 25.2, and 8.6 CFU/mL for Hist, Pre, and Test, respectively. *Escherichia coli* counts for samples collected from the front-half chiller were 6.7, 6.9, and 2.5 CFU/mL for Hist, Pre, and Test, respectively. *Salmonella* prevalence was reduced from 8.7% (Hist + Pre) to 4% (Test). Russell and Axtell (2005) reported that MON is superior to chlorine in reducing microbial populations in poultry chiller water.

21.7.7 Effect of Sodium Hypochlorite and Chlorine Dioxide

Stopforth et al. (2007) evaluated changes in APCs, total coliform counts (TCCs), *Escherichia coli* counts (*E. coli*), and *Salmonella* prevalence on poultry carcasses. In Plant A, *Salmonella* prevalence was reduced by 79% using multiple interventions, including chlorine dioxide and chlorine in the chiller. In Plant B, *Salmonella* prevalence was reduced by 91% using multiple interventions, including a TSP wash and chlorine in the chiller. In Plant C, *Salmonella* prevalence was reduced by 40% using multiple interventions, including TSP rinse and chlorine in the chiller (Stopforth et al., 2007).

21.7.8 Effect of Acidified Sodium Hypochlorite

Northcutt et al. (2008) conducted research to determine the effects of treating and reusing poultry chiller water in a commercial poultry-processing facility and the effects of the TOMCO Pathogen Management System (acidified chlorinated water) on *Salmonella* prevalence. Broiler carcasses and chiller water were obtained from a commercial processing facility that had recently installed a TOMCO system in Sections 2 and 3 of two 3-compartment chillers. Northcutt et al. (2008) found that 10 of 40 (25%) prechill carcasses were positive for *Salmonella*. After chilling, 9 of 40 (22%) carcasses were positive for *Salmonella*. These data showed that acidified chlorinated water had little effect on *Salmonella* prevalence. These data are in conflict with many real-world studies in which plants have achieved excellent *Salmonella* reduction using chlorine at pH levels below 6.5.

21.7.9 Effect of Peracetic Acid

A PAA mixture, which is a combination of PAA and hydrogen peroxide, was evaluated as an antimicrobial for use in poultry chillers (Bauermeister et al., 2008). To validate its effectiveness, PAA at 85 ppm was compared with the 30-ppm chlorine treatment in a commercial setting. Bauermeister et al. (2008) reported that, at 85 ppm, PAA reduced *Salmonella*-positive carcasses by 92% exiting the chiller, whereas treatment with 30 ppm of chlorine reduced *Salmonella* by 57%. Moreover, PAA reduced *Campylobacter* species-positive carcasses exiting the chiller by 43%, while chlorine resulted in a 13% reduction. These results suggest that PAA is an effective antimicrobial in poultry chiller applications. In another study, Bauermeister et al. (2008) evaluated PAA as an antimicrobial for use in poultry chillers. When compared to chill water containing chlorine at 30 ppm, PAA concentrations as low as 25 ppm were effective in decreasing *Salmonella* spp.

References

Bauermeister L J, McKee S R, Townsend J C, and Bowers J W J (2008), Validating the efficacy of peracetic acid mixture as an antimicrobial in poultry chillers, *Journal of Food Protection*, 71, 1119–1122.

Dickens J A, and Whittemore A D (1995), The effects of extended chilling times with acetic acid on the temperature and microbiological quality of processed poultry carcasses, *Poultry Science*, 74, 1044–1048.

Izat A L, Colberg M, Adams M H, Reiber M A, and Waldroup P W (1989), Production and processing studies to reduce the incidence of salmonellae on commercial broilers, *Journal of Food Protection*, 52, 670–673.

Lillard H S (1990), The impact of commercial processing procedures on the bacterial contamination and cross-contamination of broiler carcasses, *Journal of Food Protection*, 53, 202–204.

Northcutt J K, Ingram K D, Huezo R I, and Smith D P (2008), Microbiology of broiler carcasses and chemistry of chiller water as affected by water reuse, *Poultry Science*, 87, 1458–1463; erratum: 2008, 87(10), 2173.

Rathgeber B M, and Waldroup A L (1995), Antibacterial activity of a sodium acid pyrophosphate product in chiller water against selected bacteria on broiler carcasses, *Journal of Food Protection*, 58, 530–534.

Rice J (2006), *Salmonella* interventions in the broiler industry. http://www.fsis.usda.gov/PDF/Slides_022406_EKrushinskie.pdf.

Russell S M (2007), Chiller management key to pathogen reduction, *PoultryUSA Magazine*, March, 24, 25.

Russell S M, and Axtell S (2005), Monochloramine versus sodium hypochlorite as antimicrobial agents for reducing populations of bacteria on broiler chicken carcasses, *Journal of Food Protection*, 68, 758–763.

Stopforth J D, O'Connor R, Lopes M, Kottapalli B, Hill W E, and Samadpour M (2007), Validation of individual and multiple-sequential interventions for reduction of microbial populations during processing of poultry carcasses and parts, *Journal of Food Protection*, 70, 1393–1401.

Tamblyn K C, and Conner D E (1997), Bactericidal activity of organic acids in combination with transdermal compounds against *Salmonella* Typhimurium attached to broiler skin, *Food Microbiology*, 14, 477–484.

Tsai L S, Molyneux B T, and Schade J E (1992), Chlorination of poultry chiller water: Chlorine demand and disinfection efficiency, *Poultry Science*, 71, 188, 194–195.

Vadhanasin S, Bangtrakulnonth A, and Chidkrau T (2004), Critical control points for monitoring salmonellae reduction in Thai commercial frozen broiler processing, *Journal of Food Protection*, 67, 1480–1483.

Yang Z, Li Y, and Slavik M F (1999), Antibacterial efficacy of electrochemically activated solution for poultry spraying and chilling, *Journal of Food Science*, 64, 469–472.

Chapter 22

Proper Use of Chlorine in Poultry-Processing Plants

22.1 Introduction

Chlorine in the form of sodium hypochlorite, calcium hypochlorite tablets, or chlorine gas is by far the most commonly used carcass and equipment disinfectant in the poultry industry in the United States (Russell and Keener, 2007). The U.S. Department of Agriculture (USDA) Food Safety and Inspection Service (FSIS) allows for addition of chlorine to processing waters at levels up to 50 ppm in carcass wash applications and chiller makeup water (USDA-FSIS, 2000). The FSIS also requires that chlorinated water containing a minimum of 20 ppm available chlorine be applied to all surfaces of carcasses when the inner surfaces have been reprocessed (due to carcass contamination) other than solely by trimming (*U.S. Code of Federal Regulations*, 9 CFR 381.91).

With recent emphasis by the USDA-FSIS on further reducing *Salmonella*, poultry plants have increased their reliance on their water chlorination program in the processing plant, including prescald bird brushes, equipment rinses, inside-outside bird washers, carcass washes, and as a disinfectant during chilling. However, there remains a limited understanding of water chlorination and proper management of water chlorination in the poultry industry (Russell and Keener, 2007).

At recommended levels, hypochlorite- (chlorine derivative) based sanitizers reduce enveloped and nonenveloped viruses. Chlorine is also effective against fungi, bacteria, and algae. However, under traditional conditions of use chlorine does not affect bacterial spores. Chlorine was first used in water treatment by the municipal water treatment facilities in Chicago and Jersey City in 1908. Chlorine is used in three common forms for water treatment: elemental chlorine (chlorine gas), sodium hypochlorite (bleach) solution, and dry calcium hypochlorite pellets. The amount of hypochlorite (OCl⁻) varies depending on the type of chlorine used. One pound of Cl_2 generates the equivalent of 1 gallon of 12.5% NaOCl, and 1.5 pounds of $Ca(OCl)_2$ (65%) (water quality products).

22.2 Types of Chlorine Used in the Poultry Industry

22.2.1 Chlorine Gas

Chlorine in its elemental state is a halogen gas (Cl_2), which is highly toxic and corrosive. Because of safety concerns with chlorine gas, many food-processing facilities have changed to either sodium hypochlorite or calcium hypochlorite for water treatment.

Chlorine gas and sodium hypochlorite (NaOCl) can be produced in an electrochemical process depending on the process conditions (Equation 22.1). For NaOCl production, Cl_2 gas is passed through sodium hydroxide (NaOH) solution. The NaOH reacts with the Cl_2 and produces NaCl, NaOCl, and water as follows:

$$2 \text{ NaOH} + \text{Cl}_2 \text{ (g)} \rightleftharpoons \text{NaCl} + \text{NaOCl} + \text{H}_2\text{O} \qquad (22.1)$$

Sodium hypochlorite is the most stable and lowest-cost form of chlorine available.

22.2.2 Sodium Hypochlorite

In most food plant applications, chlorine is purchased as sodium hypochlorite (NaOCl) solution. Sodium hypochlorite solutions used in the food plant contain between 5% and 30% and sodium hypochlorite. Household bleach typically contains 5.25% NaOCl. It should be noted that household cleaners and sanitizers are not acceptable for USDA-FSIS-inspected food plants unless accepted by the USDA. Commercial forms of sodium hypochlorite are provided in a range of concentrations from 3% to 50%. The most commonly used form in poultry-processing plants is commercial bleach, which contains 12.5% NaOCl. This is the most popular form of chlorine used in poultry plants worldwide.

22.2.3 Calcium Hypochlorite

Available in granular or pellet form, calcium hypochlorite is generally more expensive to use than other hypochlorite forms. Some companies use calcium hypochlorite because they are able to control the concentration more effectively than using other forms of chlorine.

Table 22.1 Advantages and Disadvantages of Sodium Hypochlorite Use in Poultry Processing	
Advantages	*Disadvantages*
Low cost	Activity greatly influenced by pH
Familiar: proven technology	Irritating agent
Relatively nontoxic	Inactivated by organic matter
Wide germicidal activity	Less active at low temperature
Effective at low concentrations	Carcinogenic by-products
Bacteria cannot become resistant	High corrosivity
Kills bacteria in more than one way	Not accepted in Canada and Europe

Chlorine-based sanitizers are very low in cost and can control bacteria in food-processing plants when used appropriately. The advantages and disadvantages of using chlorine sanitizers are listed in Table 22.1.

22.3 Germicidal Mechanism of Action

Chlorine has a broad spectrum of activity against bacteria because its mechanism of action is so comprehensive. Chlorine compounds cause biosynthetic alterations in cellular metabolism and phospholipid destruction, formation of chloramines that interfere in cellular metabolism, oxidative action with irreversible enzymatic inactivation in bacteria, and lipid and fatty acid degradation (Estrela et al., 2002). Moreover, Camper and McFeters (1979) reported that oxygen uptake experiments showed that chlorinated cells underwent a decrease in respiration that was not immediately repaired in the presence of reducing agents (elimination of the chlorine compounds). Other reported effects of chlorine on bacteria include acidification of cell contents. Because chlorine affects bacteria in so many critical ways, bacteria are unable to develop mechanisms to become resistant to chlorine. This makes it a very effective sanitizer.

22.4 Concentration Effect

The USDA-FSIS requires a minimum of 20 ppm of chlorine in the chiller and a maximum of 50 ppm unless alternative treatments have been approved. Bacterial elimination is greatly dependent on the concentration of chlorine used and the contact time. For example, bacterial reduction will be greater in chill systems than spray systems because of the greatly increased contact time (45 minutes to 1 hour

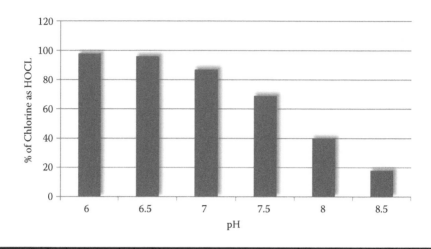

Figure 22.1 **The effect of pH on the percentage of chlorine in the hypochlorous acid state.**

vs. a couple of minutes). However, the concentration of chlorine is not nearly as important as the amount of organic material in the water in relation to the concentration of chlorine. For example, in municipal water systems, where very little organic material is present in the water, a very low level of chlorine (1 to 2 ppm) is effective for eliminating bacteria. However, in poultry chill systems, where high organic loads are encountered, often 50 ppm is not nearly sufficient.

22.5 Effect of pH on the Efficacy of Chlorine

To obtain maximum results with chlorine disinfectants, they must remain in contact with surfaces for several minutes. The pH of the water used for dilution should be below 6.5 to be most effective. Many poultry chillers operate at a pH of 7 to 8 because the water entering the plant has a neutral or high pH. Please refer to Figure 22.1, which shows the percentage of hypochlorous acid formed in waters at different pH values and at different temperatures.

22.6 Effect of Temperature and pH on Hypochlorous Acid Formation

Because hypochlorous acid is much more effective at killing bacteria than the hypochlorite ion (−OCL), the pH of chill systems and equipment sprays should be kept below 6.5. This is accomplished by adding acids such as citric acid or sodium acid sulfate or by adding carbon dioxide gas. Carbon dioxide gas may be obtained by piping gas from the CO_2 tanks used to make carbon dioxide snow for packaging. The CO_2, when in contact with water, forms carbonic acid and lowers the pH of the water. In general, the pH of the chiller should be maintained above 5 to prevent corrosion of equipment.

22.7 Effect of Organic Load

A major consideration when using chlorine as a disinfectant is that free chlorine (hypochlorous acid, hypochlorite ion, or elemental chlorine) is highly reactive and rapidly oxidizes, bleaches, or otherwise reacts with any number of substances, such as fat, blood, fecal material, or protein (Russell and Axtell, 2005). Chlorine encounters high organic loads when used in poultry-processing facilities regardless of whether it is used on carcasses in chiller systems or on equipment surfaces. Poultry process waters may have extremely high levels of total organic carbon (TOC) and a correspondingly high chemical oxygen demand (COD; Russell and Axtell, 2005). Factors such as poor bleed out, excessive scalder temperatures, poor washing of carcasses during and after evisceration, and excessive fecal caking on the surface of the chickens coming into the plant all may contribute to very high organic loading of the chiller systems. Any free chlorine added to these high-demand waters is rapidly consumed, becoming unavailable for disinfection. If the chlorine demand in these waters is not satisfied, then a free chlorine residual cannot be established. A typical poultry chiller may have a chlorine demand of 400 ppm that cannot be overcome by 50-ppm (maximum allowable by USDA) chlorine in the makeup water. Experiments conducted at the USDA Western Regional Research Center, Agricultural Research Service (ARS), concluded that a free chlorine residual could not be established in a commercial poultry chiller even by adding up to 400 ppm of free chlorine (Tsai et al., 1992). Therefore, to maximize chlorine use in poultry chillers, efforts should be made to reduce the amount of organic material in these systems. Prescald bird brushes, effective carcass rinse systems, proper bleed-out procedures, countercurrent scalders and chillers, and proper freshwater makeup in scalders and chillers all contribute to lowering organic loading of the chillers.

22.8 Disinfection By-Products

The widespread use of free available chlorine (FAC) as a disinfectant in food processing has raised food safety concerns regarding the potential for trihalomethane (THM) formation and chlorine incorporation into the food. THM compounds have been implicated as possible cancer-causing agents. Several studies have reported on incorporation of chlorine into beef, pork, chicken, and shrimp (Cunningham and Lawrence, 1977; Ghanbari et al., 1982a, 1982b; Johnston et al., 1983). Immersion of shrimp in 150-ppm chlorine for 30 minutes resulted in 2% of the FAC being incorporated into the shrimp, and 75% of this amount was detectable in the edible portion. These authors found that chlorine bound more readily to unsaturated fatty acids. The fact that poultry is high in unsaturated fatty acids, coupled with other research, such as that conducted by Kanner and Kinsella (1983), which found HOCl had the ability to destroy antioxidants, further raised the level of concern with the widespread use of FAC in the poultry-processing industry.

The USDA-FSIS allows the addition of chlorine to poultry-processing water at levels of up to 50 ppm in carcass wash applications and chiller makeup water (USDA-FSIS, 2000). If an immersion chiller is treated directly, the concentration of FAC in the chiller must be less than 50 ppm before the first carcass exits the chiller, and after the carcasses have started exiting the chiller, the FAC in the water returning from the redwater chilling system must be less than 5 ppm before that water reenters the chiller (USDA-FSIS, 2003, http://c3.org/chlorine_issues/disinfection/c3white2003.html#top).

Research has shown that an FAC residual in a loaded immersion chiller cannot be established under practical conditions (Axtell et al., 2006). One study demonstrated that the addition of 400 ppm FAC to equilibrated chiller water was insufficient in establishing an FAC residual since the chlorine demand in the chiller exceeded the 400-ppm dose (Tsai et al., 1992). The entire amount of added FAC was consumed in the chiller; some of it reacted with chiller contents such as ammonia and organic amines from the chicken carcasses to randomly form various chlorinous compounds, such as the beneficial biocide monochloramine and undesired nonbiocidal organic chloramines (Axtell et al., 2006). Since none of the added FAC remained as free chlorine, the total chlorine content of the chiller was comprised entirely of combined chlorine (Axtell et al., 2006). The use of FAC as a chiller chlorination control metric does not capture the combined chlorine component of the water.

22.9 Odor Problems

Some poultry processors have experienced a problem with gassing off of chlorine-containing compounds, causing the USDA-FSIS to shut the line down. This problem has been closely linked to the addition of ammonia to the water by municipal water treatment facilities to prevent scale buildup. Ammonia in the water, on contact with the chlorine added by the processor, can form di- and trichloramines (nitrogen di- or trichloride). These compounds gas off and are objectionable, causing burning of the eyes and throat. Even when chlorine is added at very low concentrations (20 ppm), the USDA-FSIS may still require plants to stop the line because of the off odors. This is a city water problem and not a plant problem.

22.10 Proper Chlorine Management

22.10.1 Equipment Sprays or Carcass Rinses

The chlorine concentration in equipment sprays and carcass rinses should be maintained at 40 to 45 ppm. The pH of the chlorinated water should be monitored, and if the pH is above 6.5, citric acid or CO_2 should be added to ensure that the final pH of the water with the bleach added is below 6.5. Bleach tends to increase the pH of water slightly. Interestingly, scientists at the USDA-ARS published a study that demonstrated that using 50-ppm chlorine in an inside-outside bird washer had no measurable impact on *Salmonella* populations on chicken carcasses (Smith et al., 2004). This is interesting in light of how often the industry relies on chlorinated carcass sprays as interventions.

22.10.2 Poultry Chillers

As mentioned in Chapter 21, significant *Salmonella* reduction (in both numbers and prevalence) can be more readily accomplished in a properly balanced chiller than anywhere else in the processing plant. Most studies demonstrated that the chiller can significantly reduce *Salmonella* prevalence (Izat et al., 1989) if operating properly. As mentioned previously, the pH of the chiller should be maintained below 6.5 for optimal results and to ensure that chlorine is maintained in the hypochlorous acid form. Moreover, the maintenance of a high level of fresh makeup water and a countercurrent directional flow is essential. In addition, the concentration of chlorine should be maintained at or about 50 ppm in the makeup water, and the concentration of organic material (digesta, fat, blood) should be kept as low as possible for the chlorine in the chiller to do its job.

Chlorine has been and continues to be an effective means of reducing bacteria on equipment surfaces and poultry carcasses. External pressures such as concern about the formation of carcinogens, gassing off and employee safety, the effect of organic loading on the efficacy, and international restrictions may spell the eventual demise of this disinfectant. However, because of its low cost, ease of use, and availability, it is still the most widely used disinfectant in the industry.

References

Axtell S P, Russell S M, and Berman E (2006), Effect of immersion chilling of broiler chicken carcasses in monochloramine on lipid oxidation and halogenated residual compound formation, *Journal of Food Protection*, 69, 907–911.

Camper A K, and McFeters G A (1979), Chlorine injury and the enumeration of waterborne coliform bacteria, *Applied and Environmental Microbiology*, 37, 633–641.

Cunningham H M, and Lawrence G L (1977), Effect of exposure of meat and poultry to chlorinated water on the retention of chlorinated compounds and water, *Journal of Food Science*, 42, 1504–1505, 1509.

Estrela C I, Estrela C R A, Barbin E L, Spano J C E, Marchesan M A, and Pecora J D (2002), Mechanism of action of sodium hypochlorite, *Brazilian Journal of Dentistry*, 13, 113–117.

Ghanbari H A, Wheeler W B, and Kirk J R (1982a), The fate of hypochlorous acid during shrimp processing: A model system, *Journal of Food Science*, 47, 185–187, 192.

Ghanbari H A, Wheeler W B, and Kirk J R (1982b), Reactions of aqueous chlorine and chlorine dioxide with lipids: Chlorine incorporation, *Journal of Food Science*, 47, 482–485.

Izat A L, Colberg M, Adams M H, Reiber M A, and Waldroup P W (1989), Production and processing studies to reduce the incidence of salmonellae on commercial broilers, *Journal of Food Protection*, 52, 670–673.

Johnston J J, Ghanbari H A, Wheeler W B, and Kirk J R (1983), Chlorine incorporation in shrimp, *Journal of Food Science*, 48, 668–670.

Kanner J, and Kinsella J E (1983), Lipid deterioration initiated by phagocytic cells in muscle foods: b-Carotene destruction by a myeloperoxidase-hydrogen peroxide-halide system, *Journal of Agricultural Food Chemistry*, 31, 370–376.

Russell S M, and Axtell S (2005), The effect of monochloramine versus chlorine on pathogenic, indicator, and spoilage bacteria associated with broiler chicken carcasses: A model, pilot scale, and industrial study, *Journal of Food Protection*, 68, 758–763.

Russell S M, and Keener K M (2007), Chlorine: Misunderstood pathogen reduction tool, *PoultryUSA Magazine*, May, 22–24, 26, 27.

Smith D P, Northcutt J K, and Musgrove M T (2004), Effect of commercial inside-outside bird washer (IOBW) on *Campylobacter*, *Salmonella*, *E. coli*, and aerobic plate counts (APC) of uncontaminated, contaminated, and cross-contaminated broiler carcasses, *Poultry Science*, 83(Suppl. 1), 155.

Tsai L S, Molyneux B T, and Schade J E (1992), Chlorination of poultry chiller water: Chlorine demand and disinfection efficiency, *Poultry Science*, 71, 188, 194–195.

U.S. Code of Federal Regulations, 9 CFR 381.91, 2009, Poultry Products Inspection Regulations.

U.S. Department of Agriculture, Food Safety and Inspection Service (2000), *Sanitation Performance Standards*, Directive 11,000.1, U.S. Department of Agriculture, Food Safety and Inspection Service, Washington, DC.

U.S. Department of Agriculture, Food Safety and Inspection Service (2003), *Use of Chlorine to Treat Poultry Chiller Water*, FSIS Notice 45-03, U.S. Department of Agriculture, Food Safety and Inspection Service, Washington, DC. http://c3.org/chlorine_issues/ disinfection/c3white2003.html#top; Water Quality Products. A supplement to water quality products disinfection. www.scrantongillette.com/email_images/ waterdisinfection_0208.pdf, Arlington Heights, IL.

Chapter 23

Effect of Air Chilling on *Salmonella*

23.1 Introduction

Air chilling (Figure 23.1) is a process in which carcasses are suspended on racks that are placed into a cooler or carcasses are suspended on shackles that move through a cooler until the internal temperature is below 5°C. The advantage to air chilling is that, unlike an improperly controlled chiller that enables *Salmonella* to be washed from positive carcasses to negative ones, it is more difficult for rapid air movement over the carcass to cause *Salmonella* transfer. However, the major disadvantage is that, with air chillers, no chemicals may be used to reduce *Salmonella* prevalence on carcasses as they may be in properly controlled immersion chillers.

23.2 Effect of Air Chilling on Bacterial Numbers

In 1994, Abu-Ruwaida et al. reported that no substantial change occurred in bacteria levels during air chilling, packaging, and cold storage; however, the finished product was heavily contaminated. In the freshly processed carcasses, mean counts ($\log_{10}$ colony-forming units [CFU]/mL) of aerobic bacteria Enterobacteriaceae, coliforms, *Escherichia coli*, *Campylobacter*, and *Staphylococcus aureus*, respectively, were 6.6, 4.5, 4.1, 3.6, 5.2, and 2.7 on the first sampling day, and 6.5, 4.6, 4.9, 3.6, 4.7, and 4.1 on the second day. *Salmonella* was present in all birds examined,

Figure 23.1   A poultry air chiller.

including those coming directly from the farm. This study demonstrated that air chilling had no impact on *Salmonella* on poultry (Abu-Ruwaida et al., 1994).

Carraminana et al. (1997) conducted a *Salmonella* spp. survey at 11 sampling sites in a poultry slaughter establishment in Spain. *Salmonella* prevalence rates increased from 30% in fecal material collected from incoming birds to 60% in air-chilled carcasses (Carraminana et al., 1997). This study showed the problem with trying to control *Salmonella* on carcasses when using air chilling.

The bacterial loads in the air at different locations of a poultry-slaughtering and -processing plant were examined (Ellerbroek, 1997). Enterobacteriaceae were detected at 2.02 $\log_{10}$ CFU/m^3 in the air-chilling room and 2.06 $\log_{10}$ CFU/m^3 in the spray-chilling room (Ellerbroek, 1997). The data collected in this study indicated that high levels of enteric bacteria were detected in the air-chilling and spray-chilling areas, which means that *Salmonella* would be able to survive in these environments.

Mead et al. (2000) studied cross contamination of bacteria during air chilling of poultry carcasses using a nalidixic acid-resistant strain of *Escherichia coli* K12 as a marker organism. The authors reported that, despite the ease of microbial transmission from inoculated carcasses, cross contamination during air chilling is likely to be less than that occurring at earlier stages of poultry processing, when carcasses are more heavily contaminated.

In 2002, Sanchez et al. compared the microbiological loads and the prevalence of *Salmonella* spp. and *Campylobacter* spp. on broiler carcasses subjected to immersion chilling and air chilling. The results of this study indicated that the prevalence of *Salmonella* spp. and *Campylobacter* spp. tends to be significantly lower in air-chilled broilers, suggesting that cross contamination may be more prevalent for immersion-chilled broilers. The authors reported that the chilling method used during processing may influence the microbial profile of postchilled broilers. This study was conducted with an immersion chiller that was not properly controlled. *Salmonella* prevalence is generally lower for immersion chillers when the chillers are balanced and operating properly.

Fluckey et al. (2003) studied the microbiological profile of an air-chilling poultry process through the processing plant. Generic *E. coli* counts were reduced from 3.08 to 2.20 $\log_{10}$ CFU/mL of rinse for after evisceration to after air chilling samples, respectively. These data indicate that air chilling was able to slightly reduce *E. coli* counts by 0.88 $\log_{10}$ CFU/mL. The authors found no reductions in numbers of *Campylobacter* or *Salmonella* during air chilling, indicating that air chilling in this study was an ineffective intervention.

In 2008, Berrang et al. compared the effect of two chilling methods on broiler carcass bacteria. Broiler carcasses were cut in half along the dorsal-ventral midline; one of the halves was subjected to an ice water immersion chill in an agitated bath for 50 minutes; the reciprocal half was subjected to an air chill in a 1°C cold room for 150 minutes. Total aerobic bacteria, coliforms, *Escherichia coli*, and *Campylobacter* were enumerated from half-carcass rinses. The authors reported that their data showed that immersion-chilled carcasses had lower numbers of bacteria; however, the difference was not large and may have been due to simple dilution (Berrang et al., 2008).

In the research literature, there are numerous conflicting studies with regard to the antimicrobial efficacy of air versus immersion chilling. With air chilling, there is no real opportunity to use chemical intervention. In most countries where air chilling is used, chemical intervention is not allowed. If an immersion chilling system is operated properly, it is a very effective intervention strategy for eliminating *Salmonella*. If it is not being operated properly, *Salmonella* may spread from carcass to carcass, increasing prevalence.

References

Abu-Ruwaida A S, Sawaya W N, Dashti B H, Murad M, and Al-Othman H A (1994), Microbiological quality of broilers during processing in a modern commercial slaughterhouse in Kuwait, *Journal of Food Protection*, 57, 887–892.

Berrang M E, Zhuang H, Smith D P, and Meinersmann R J (2008), The effect of chilling in cold air or ice water on the microbiological quality of broiler carcasses and the population of *Campylobacter*, *Poultry Science*, 87, 992–998.

Carraminana J J, Yanguela J, Blanco D, Rota C, Agustin A I, Arino A, and Herrera A (1997), *Salmonella* incidence and distribution of serotypes throughout processing in a Spanish poultry slaughterhouse, *Journal of Food Protection*, 60, 1312–1317.

Ellerbroek L (1997), Airborne microflora in poultry slaughtering establishments, *Food Microbiology*, 14, 527–531.

Fluckey W M, Sanchez M X, Mckee S R, Smith D, Pendleton E, and Brashears M M (2003), Establishment of a microbiological profile for an air-chilling poultry operation in the United States, *Journal of Food Protection*, 66, 272–279.

Mead G C, Allen V M, Burton C H, and Corry J E L (2000), Microbial cross-contamination during air chilling of poultry, *British Poultry Science*, 41, 158–162.

Sanchez M X, Fluckey W M, Brashears M M, and Mckee S R (2002), Microbial profile and antibiotic susceptibility of *Campylobacter* spp. and *Salmonella* spp. in broilers processed in air-chilled and immersion-chilled environments, *Journal of Food Protection*, 65, 948–956.

Chapter 24

Postchill Processes: Dips and Sprays

24.1 Introduction

Poultry processors are employing the "hurdle hypothesis" to reduce *Salmonella* at different locations throughout the plant. The hurdle hypothesis is the premise that the more hurdles (i.e., interventions) that are employed against *Salmonella*, the less likely it is that *Salmonella* cells will be able to survive until the end of the process. As a final intervention and hurdle, companies are now using postchill dips or sprays (Russell, 2007).

These systems are advantageous in that the chickens are as clean as they will be throughout the process, and the ability of any given chemical to contact bacteria on the surface of the skin without interference from organic material is highest at this point because of air agitation and movement in the immersion chiller. In general, fecal material, fat, protein, blood, bile, and bacterial biofilms that may be on the surface of the carcass have been removed by the time the carcass exits the chiller. Thus, the bacteria are most susceptible to disinfection at this point.

24.2 Chemicals Used in Postchill Dips

The chemicals that are being used for postchill dips include acidified sodium chlorite (ASC); hypochlorous acid; peracetic acid; cetylpyridinium chloride; mixtures of organic or inorganic acids, including citric, lactic, acetic, phosphoric, sulfuric, and hydrochloric acids; chlorine dioxide; and electrolyzed oxidative acidic water, to name a few. The U.S. Poultry and Egg Association industry survey by Rice (2006) indicated that, of the companies that participated in the survey, 67% of the industry used ASC, 25% used chlorine dioxide (this has been reduced dramatically since the survey), and 8% used hypochlorous acid. The dip tanks used for these applications generally vary from small 50- to 100-gallon tanks to much larger (5,000- to 10,000-gallon) prechiller-type tanks. Likewise, the contact time used by these poultry companies varies from 8 seconds to 30 minutes. The spray systems are generally similar to those used for online reprocessing. Not many studies have been published regarding postchill dip systems. ASC was investigated by Sexton et al. (2007).

24.3 Effect of Acidified Sodium Chlorite

Sexton et al. (2007) conducted a trial on the efficacy of ASC on *Salmonella* and *Campylobacter* on chicken carcasses after they exited the screw chiller of a commercial plant. On untreated carcasses, mean $\log_{10}$ total viable count was $2.78/cm^2$ compared with $1.23/cm^2$ on treated carcasses. Prevalence of *Escherichia coli*, *Salmonella*, and *Campylobacter* was 100%, 90%, and 100%, respectively, on untreated carcasses and 13%, 10%, and 23%, respectively, on treated carcasses. The significant reductions in prevalence demonstrated that ASC is an effective postchill intervention option (Sexton et al., 2007).

Figures 24.1, 24.2, and 24.3 depict aerobic plate counts (APCs), *E. coli*, and *Salmonella* reductions observed at two different poultry-processing plants. Figures 24.1 and 24.2 show reductions in APC or *E. coli* using a mixture of sulfuric acid, ammonium sulfate, and copper sulfate for 10 seconds. Figure 24.3 shows reductions in *Salmonella* prevalence using ASC for 8 seconds.

Because the organic loading is so low on carcasses after chilling, the efficacy of oxidant-type chemicals at this point in the process is high. These chemicals are most effective when they can directly contact bacteria without interference from organic material. Likewise, acid-based sanitizers are effective because they are able to have an extended contact time. Whether a spray or dip is used, if the chemical is applied after the chiller, no other water washes are used. Therefore, the contact time may be hours or days. Chemicals that leave a residual may be effective in postchill dips or sprays, and they may significantly extend the shelf life of the product because of this residual; however, the U.S. Food and Drug Administration (FDA) requires that chemicals that have a "material effect" on the product after packaging, such as extending shelf life, must be added to the label as an additive (preservative). Thus, the processor would have to add the chemical to the label, and this is generally viewed in a negative light in terms of consumer acceptance.

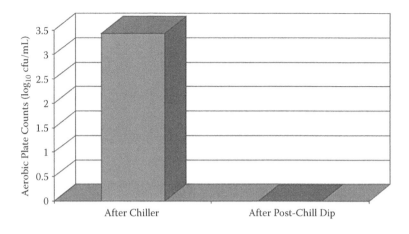

Figure 24.1 **APC counts on broiler carcasses before or after exposure to a postchill dip tank containing sulfuric acid, ammonium sulfate, and copper sulfate at a pH of 2.0 for 10 seconds.**

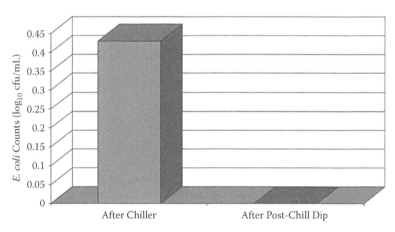

Figure 24.2 *Escherichia coli* **counts on broiler carcasses before or after exposure to a postchill dip tank containing sulfuric acid, ammonium sulfate, and copper sulfate at a pH of 2.0 for 10 seconds.**

The decision to use a dip or spray system to apply the chemicals postchill is based on whether the company normally rehangs the carcasses postchill (Russell, 2007). For example, for companies that normally process whole ready-to-cook carcasses, these carcasses are packaged after chilling and would not be rehung. Thus, a dip system is more useful in this situation. If the company debones most or all of the carcasses, then the carcasses will be rehung on a line, and the spray system may be much easier to install and use in this scenario.

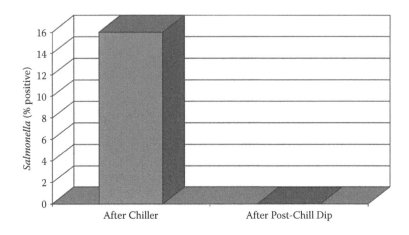

Figure 24.3 *Salmonella* **prevalence on broiler carcasses before or after exposure to a postchill dip tank containing acidified sodium chlorite at 1,200 ppm for 8 seconds.**

Overall, poultry companies are achieving success using postchill dips or sprays and are finding that the interventions throughout the plant combined with a postchill dip system can be effective for reducing *Salmonella* and other bacteria to acceptable levels.

References

Rice J (2006), *Salmonella* interventions in the broiler industry. http://www.fsis.usda.gov/PDF/Slides_022406_EKrushinskie.pdf.

Russell S M (2007), Using post-chill dips and sprays, *PoultryUSA Magazine*, March, 26.

Sexton M, Raven G, Holds G, Pointon A, Kiermeier A, and Sumner J (2007), Effect of acidified sodium chlorite treatment on chicken carcasses processed in South Australia, *International Journal of Food Microbiology*, 30, 252–255.

Chapter 25

Other Novel Approaches to Elimination of *Salmonella* on Carcasses

25.1 Introduction

Use of bacteriophages to eliminate *Salmonella* on live birds was discussed in Chapter 9, Section 9.5. However, this approach has also been used on chicken skin.

25.2 Effect of Lytic Bacteriophages on *Salmonella* and *Campylobacter* on Chicken Skin

In 2003, Goode et al. studied the effect of lytic bacteriophages (Figures 25.1 and 25.2) applied to chicken skin that had been experimentally contaminated with *Salmonella enterica* serovar Enteritidis (SE) or *Campylobacter jejuni*. Phages rapidly reduced the recoverable bacterial numbers by up to 2 $\log_{10}$ colony-forming units (CFU)/mL over 48 hours (Goode et al., 2003). It is important to note that samples to be tested for *Salmonella* must be collected from carcasses by inspectors from the U.S. Department of Agriculture, Food Safety Inspection Service (USDA-FSIS), within 1.5 hours of slaughter. These samples are tested within 24 hours of collection. Thus, the total time from application of the phages to the testing of the sample

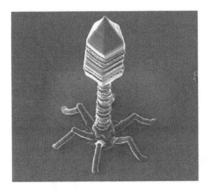

Figure 25.1 Ion micrograph of a bacteriophage. (From http://www.zyvexlabs. com/EIPBNuG/2005MicroGraph.html.)

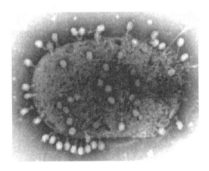

Figure 25.2 Bacteriophage attacking an *E. coli* bacterium. (Figure used with permission from Dr. M. V. Parthasarathy. http://www.bacteriophagetherapy. info/ECF40946-8E2F-4890-9CA6-D390A26E39C1/9D10A630-EEB1-46C5-BEFD-04883AAD7C2A.html.)

is only 25.5 hours, which is far less than the time used in this study (48 hours). It is likely that *Salmonella* and *Campylobacter* numbers were reduced in less time than used for this study, and this methodology would be effective in practical application. Bielke et al. (2007) studied the effect of bacteriophages targeted against *Salmonella* and applied to poultry carcasses. Fresh-processed chicken carcasses were inoculated with either SE or *Salmonella* Typhimurium (ST) and sprayed with 5 mL of bacteriophage. The authors found that the bacteriophages significantly reduced the recovery of SE. No SE was detected in two trials, and greater than 70% reduction was seen in the other two trials. ST was also significantly reduced in the two trials in which it was included. Bielke et al. (2007) concluded that their experiments suggested that bacteriophages could be an inexpensive and safe method for the reduction of *Salmonella* on broiler carcasses.

References

Bielke L R, Hargis B M, Tellez G, Donoghue D J, Higgins S E, and Donoghue A M (2007), Use of wide-host-range bacteriophages to reduce *Salmonella* on poultry products, *International Journal of Poultry Science*, 6, 754–757.

Goode D, Allen V M, and Barrow P A (2003), Reduction of experimental *Salmonella* and *Campylobacter* contamination of chicken skin by application of lytic bacteriophages, *Applied and Environmental Microbiology*, 69, 5032–5036.

Chapter 26

Biomapping *Salmonella* on Broiler Carcasses in Poultry-Processing Plants: Case Studies

26.1 Introduction

One of the first questions to be asked if the U.S. Department of Agriculture, Food Safety Inspection Service (USDA-FSIS), indicates that a processor is failing the *Salmonella* performance standard is, What happens to *Salmonella* levels on broiler carcasses as they progress through each intervention? (Russell, 2007). Most of the time, the plant managers have no idea. This is why it is essential that *Salmonella* biomapping be done by every plant on a regular basis. This information is essential to understand which processes are performing effectively for reducing *Salmonella* and which ones need to be adjusted. Without biomapping information, tuning the plant for *Salmonella* reduction is impossible.

Reducing *Salmonella* on finished carcasses requires a comprehensive, multihurdle approach. The point of biomapping is that if the plant is tuned correctly, the plant will always keep *Salmonella* levels below the performance standard, regardless of the incoming load on chickens. No individual procedure or step is adequate to accomplish this task (i.e., there is no silver bullet). The USDA stated that "intervention strategies aimed at reducing fecal contamination and other sources of *Salmonella* on raw product should be effective against other pathogens" (USDA, 1996). This statement is misleading in that reducing fecal contamination has not been sufficient to reduce *Salmonella* prevalence. Reducing fecal contamination may be effective for reducing the number of *Salmonella* on each carcass; however, only one *Salmonella* cell is required on a carcass to produce a "positive" result. Thus, unless all *Salmonella* are eliminated, the carcass will remain positive (Russell, 2007).

26.2 Data Collection Methods

To begin the biomapping process, data must be collected. Carcass samples should be collected at the following locations to determine which interventions are working:

1. Post-bleed out
2. Postscald
3. Pre-online reprocessing (OLR)
4. Post-OLR
5. Postchill
6. Post-postchill dip

Aerobic plate counts, *Escherichia coli* counts, and *Salmonella* prevalence should be determined for carcasses at each of these locations over a period of time. For example, an appropriate sample size would be to collect 96 carcasses at each sampling point over a period of 12 days. This sample size is sufficient to determine variability between flocks and if the interventions are consistent over time.

26.3 Real-World Biomaps

The following are real-world examples and represent data collected from various processors nationwide. Due to the confidential nature of this subject, none of the companies has been identified.

The biomap of Figure 26.1 shows the negative impact that picking has on *Salmonella* prevalence. Often, *Salmonella* are spread to negative carcasses during picking. In some cases, the prevalence can exceed incoming load.

The biomap in Figure 26.2 clearly indicates that the scalder needs to be adjusted to ensure that *Salmonella* prevalence is reduced during scalding. This may be done with an increase in temperature, increase in freshwater input, or input of a chemical sanitizer, such as an acid. Moreover, the OLR in this plant is functioning well and should not be adjusted.

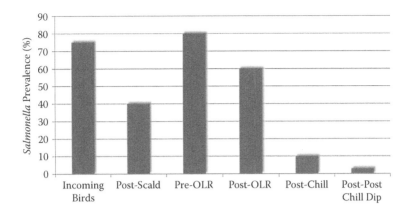

Figure 26.1   Typical poultry-processing plant biomap showing *Salmonella* prevalence as the carcasses progress down the processing line.

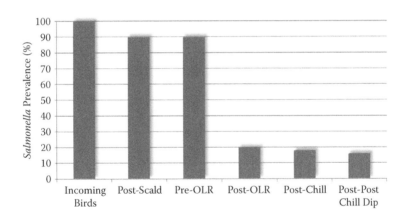

Figure 26.2   Biomap 2 showing *Salmonella* prevalence as the carcasses progress down the processing line, demonstrating that in this plant, the scalder is ineffective, and the OLR is the only effective intervention.

The biomap profile in Figure 26.3 demonstrates that the breeding, hatching, and grow-out operations are holding *Salmonella* to low levels on incoming birds. However, the scalder is not improving the *Salmonella* profile at all and should be adjusted. The OLR system is ineffective, and the plant is spending its money on this expensive chemical treatment but receiving no beneficial impact on *Salmonella* prevalence.

A biomap with the profile of Figure 26.4 demonstrates that the breeding, hatching, and grow-out operations are holding *Salmonella* to low levels on incoming

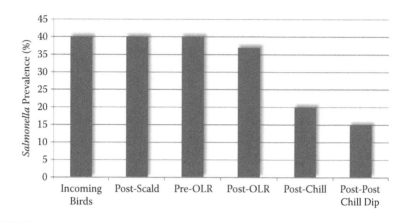

Figure 26.3   Biomap 3 showing *Salmonella* prevalence as the carcasses progress down the processing line, demonstrating that in this plant, incoming *Salmonella* load is low, and the scalder and OLR are not effective.

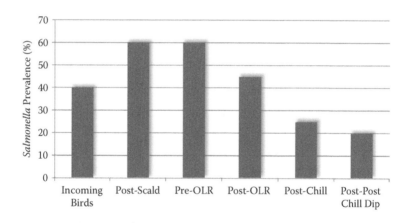

Figure 26.4   Biomap 4 showing *Salmonella* prevalence as the carcasses progress down the processing line, demonstrating that in this plant, the scalder is inoculating the carcasses with *Salmonella,* and the postchill dip is ineffective.

birds. However, the scalder is being run at too low a temperature and is inoculating the carcasses as they progress through the system. The postchill dip system is not effective, and the company is wasting its money on this expensive chemical treatment and is receiving no beneficial impact on *Salmonella* prevalence.

A biomap with the profile in Figure 26.5 demonstrates that the breeding, hatching, and grow-out operations are holding *Salmonella* to low levels on incoming birds. However, the scalder is being run at too low a temperature and is inoculating the carcasses as they progress through the system. The chiller and postchill

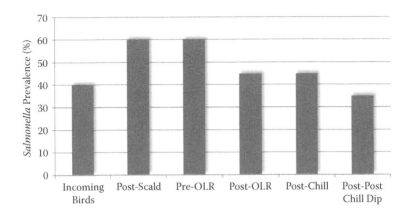

Figure 26.5 Biomap 5 showing *Salmonella* prevalence as the carcasses progress down the processing line, demonstrating that in this plant, the scalder is inoculating the chickens with *Salmonella,* and the chiller and postchill dip are ineffective.

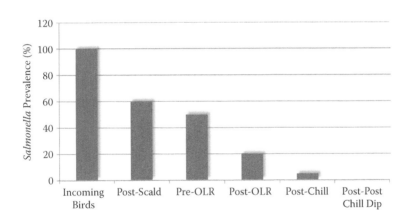

Figure 26.6 Biomap 6 showing *Salmonella* prevalence as the carcasses progress down the processing line, demonstrating that in this plant, each intervention is effective.

dip system are not effective, and the company is not able to meet the *Salmonella* performance standard. Thus, the scalder, chiller, and postchill dip system should be adjusted to perform correctly.

In Figure 26.6, the incoming birds are highly contaminated, but the scalder is balanced, the pickers are not contributing to the problem, the OLR system is working well, and the chiller and postchill dip are balanced appropriately. This scenario is achievable. A handful of plants around the country are able to operate as depicted in this figure.

26.4 Violation of the *Salmonella* Performance Standard

Every poultry-processing plant is different. When a problem occurs, each plant must be evaluated, and a unique plan should be developed to correct the problem. The good news is that problems may be identified and corrected provided enough information is available. Often, when working on these problems, the information is not available and must be collected after the problem occurs—in some cases, after the plant has been shut down. Thus, a catch 22 occurs. The plant management asks the following question: How can we demonstrate that we have fixed the *Salmonella* problem when we cannot run the plant?

The examples presented are but a few of the possibilities that may be seen in the industry. Sometimes, the problem is simple; at other times, multiple fixes are required. On some occasions, the plant cannot be balanced, and efforts must be made in the field. This occurs when the plant is using very old equipment, is unable to install new equipment due to cost or space limitations, or may not implement certain chemicals due to poor ventilation or export regulations. In general, it is much more expensive to implement field interventions than to correct the problems in the plant.

Every plant should strive to achieve a pattern as presented in Figure 26.6. This level of performance is achievable. Unfortunately, some companies choose to wait until the last USDA-FSIS *Salmonella* testing set failure (Set C) before they seek help. At this point, enormous amounts of money are required to fix the problem. Moreover, once the USDA-FSIS begins to see *Salmonella* set failures, the plant is usually placed under intense scrutiny. It is much less expensive and more effective to be proactive and to control the problem.

References

Russell S M (2007), Troubleshooting *Salmonella* issues in poultry processing facilities, *Zootechnica World's Poultry Journal*. http://www.zootecnicainternational.com/article-archive/processing/693-troubleshooting-salmonella-issues-in-poultry-processing-facilities-.html.

U.S. Department of Agriculture, Food Safety Inspection Service (1996), Pathogen reduction; hazard analysis and critical control point (HACCP) systems, *Federal Regulation*, 61, 38806–38989.

Chapter 27

Verification of the Efficacy of Intervention Strategies

27.1 Introduction

A study was conducted to address concerns by the Food Safety Inspection Service (FSIS) of the U.S. Department of Agriculture (USDA) regarding the control of physical, chemical, and biological hazards that were identified by a large poultry-processing facility (Russell, 2005). The intent of the study was to conduct a series of research studies developed using statistically derived experimental designs to evaluate if each of the interventions used in the poultry-processing operation were adequate to address the hazards listed in the hazard analysis and critical control point (HACCP) program and to verify that the plant could control bacterial populations on poultry. The USDA-FSIS will be requiring that more processing facilities conduct verifications on their intervention strategies in the future. This chapter serves as a guideline for processors facing the prospect of having to prove that their interventions are working. This is an actual case study in a processing facility. The methods used, results achieved, and a full discussion of their meaning are presented. The report from which this chapter was constructed was considered satisfactory by the USDA-FSIS for verifying the efficacy of the intervention strategies in this processing plant.

27.2 Experimental Methods Used to Evaluate the Individual Interventions

The methods to evaluate the individual interventions used have been separated into specific sections based on the location within the plant or the particular intervention tested. The number of samples required to evaluate the efficacy for each of the interventions was mathematically determined by the Department of Statistics at the University of Georgia.

27.2.1 Incoming Birds

An experiment was conducted to evaluate the aerobic plate count (APC), *Escherichia coli* count, and *Salmonella* prevalence on birds coming into the processing plant. The number of chickens tested was based on variance within flocks (individual grow-out houses) to represent an appropriate number of grow-out facilities properly.

27.2.1.1 Experimental Procedure

Four carcasses from two separate flocks processed each day (the third flock on the first shift and the third flock on the second shift) were removed from the line just after bleed out using the following technique to ensure that no bias was introduced: These flocks were chosen to prevent selection of carcasses from the very beginning of a shift when process waters are clean and do not represent typical processing situations. A carcass was selected visually on the line after bleed out, then the next five carcasses were counted aloud, and the sixth carcass was selected for testing. The individual selecting the carcasses was wearing sterile examination gloves. In this way, no visual cues were used to introduce bias. The four carcasses were plugged using unscented tampons, and the necks were tied off using zip ties. This was done so that the variable of fecal or ingesta leakage was avoided. It is impossible to predict how much fecal material or ingesta will leak from bird to bird during rinsing. Therefore, it is best to avoid this variable by making sure no leakage occurs. This has been detailed in a research study by Musgrove et al. (1997). The carcasses were then individually bagged in sterile polyethylene bags and rinsed using 400 mL of sterile buffered peptone solution by conducting the whole-carcass rinse method as employed by the USDA inspectors in processing facilities. The rinsate was encoded using a three-digit number (to prevent identification by laboratory employees and the introduction of bias) and sent to the laboratory for evaluation for APC, *E. coli* counts, and *Salmonella* prevalence. The microbiological methods used are detailed in the microbiological methods section of this report. A total of 96 carcasses were evaluated over 12 days and 24 flocks. This allowed the plant to determine the level of APC, *E. coli,* and *Salmonella* coming in on the live birds. The number of carcasses tested was based on statistical models to ensure a representative number over the entire grow-out process. Incoming bacterial counts and prevalence were evaluated so that a comparison to fully processed, ready-to-cook carcasses could be made.

27.2.2 Scalder Water

Regulations in 9 CFR 416.2 (g)(3) state that water that has contacted raw product may be reused for the same purpose or upstream on the processing line provided that measures are taken to reduce physical, chemical, and microbiological contamination or adulteration of the product. This has recently been applied by the USDA-FSIS to the use of countercurrent scalders and chillers. Poultry processors must now ensure that they are addressing the physical, chemical, and biological hazards associated with the scalder water and recirculated chiller water (redwater) to meet the HACCP regulation requirements. To ensure compliance with this regulation, the plant conducted a test to make sure that the physical, chemical, and biological hazards associated with scald water were addressed, and that these hazards were lower at the exit end of the scalder than they were at the entrance end of the scalder. In particular, the biological hazard is the hazard of greatest concern. There is no physical hazard associated with the scalder. While the presence of turbidity or suspended solids does represent a hazard in that it can protect pathogenic bacteria such as *Salmonella* and enable them to survive, it is not a physical hazard. Instead, it contributes to a biological hazard. For example, no one has ever choked to death on turbidity, and no one has ever broken a tooth on suspended solids. Thus, the hazard that turbidity and suspended solids influence (biological) were assessed in this study.

27.2.2.1 Experimental Procedure

The scalder was a true countercurrent scalder in that the fresh makeup water was added at the carcass exit. Scalder water samples were collected from the carcass entrance end and the carcass exit end just prior to changeover to a new flock of chickens from the field. This was done eight times per day (four samples at each point for the third flock on the first shift and four samples at each point for the third flock on the second shift) for 12 days to represent 24 flocks of chickens. APC, *E. coli* count, and *Salmonella* prevalence were determined for each of these samples. Based on published papers, we expected that these data would provide evidence that the scalder water from which the carcasses exit would have significantly fewer indicator and pathogenic bacteria than the scalder water from the entrance end of the scalder. In this way, the company may verify that the use of scalder water in a countercurrent system is meeting the requirements of 9 CFR 416.2 (g)(3), and that the biological hazard is being reduced. The microbiological methods used for this portion of the study are described in the "Microbiological Methods" section.

27.2.3 Pre- and Postscald Carcasses

An experiment was conducted to determine if the scalder was able to reduce indicator and pathogenic bacteria on broiler carcasses. The research literature indicated that scalder temperature and dilution rate may have a significant impact on bacterial levels on chicken carcasses.

27.2.3.1 Experimental Procedure

Four carcasses from two separate flocks processed each day (the third flock on the first shift and the third flock on the second shift) were removed from the line just after bleed out, and the same number at the same times were collected from the line just after the final scald tank and just prior to picking using the aforementioned technique to ensure that no bias was introduced. The four carcasses were plugged using unscented tampons, and the necks were tied off using zip ties. The carcasses were individually bagged in sterile polyethylene bags and rinsed using 400 mL of sterile buffered peptone solution by conducting whole-carcass rinses. The rinsate was encoded as previously described and sent to the laboratory for evaluation for APC, *E. coli* counts, and *Salmonella* prevalence. The microbiological methods used are detailed in the "Microbiological Methods" section of this chapter. There were a total of 192 carcasses evaluated over 12 days and 24 flocks. This allowed the plant to determine the level of APC, *E. coli* count, and *Salmonella* prevalence on carcasses prescald and postscald. The numbers of carcasses tested were determined based on statistical models to ensure a representative sampling of the grow-out process.

27.2.4 Salvage

The salvage process is of concern because the carcasses that were sent to salvage, manually reprocessed, and cut into parts (i.e., wings, tenders, boneless breast, leg quarters) were not being treated by the online reprocessing (OLR) system. Therefore, it was necessary to verify that reprocessing methods were sufficient to control indicator populations of bacteria and *Salmonella* prevalence. The establishment followed procedures as set forth in QAP 120.00 "Reprocessing Guidelines." Until proper validation was completed and accepted by FSIS, all salvage parts that were not run through the OLR disinfection system were condemned.

27.2.4.1 Experimental Procedure

Four carcasses from two separate flocks processed each day (the third flock on the first shift and the third flock on the second shift) were removed from the line just after the USDA inspector hung it on the salvage line, and four carcasses were collected just after the carcasses were sent to the salvage holding section (after being cleaned and sanitized). The carcasses were then individually bagged in sterile polyethylene bags and rinsed using 400 mL of sterile buffered peptone solution by conducting whole-carcass rinses. The rinsate was encoded as previously described and sent to the laboratory for evaluation for APC, *E. coli* counts, and *Salmonella* prevalence. The microbiological methods used are detailed in the "Microbiological Methods" section of this chapter. There were 192 carcasses evaluated over 12 days and 24 flocks. This allowed the plant to determine the level of APC, *E. coli* counts, and *Salmonella* prevalence on carcasses presalvage and postsalvage. The number of carcasses tested was based on statistical models to ensure a representative sampling of the grow-out process.

27.2.5 Inside-Outside Bird Washer and Final Rinse

The inside-outside bird washer (IOBW) and final bird rinse have traditionally been used to remove any fecal material or ingesta that may incidentally have contacted the carcass during evisceration or remained on the carcass through scalding and picking. The IOBW and final rinse combined were evaluated to ensure that this intervention step was successfully reducing fecal or ingesta contamination.

27.2.5.1 Experimental Procedures

Ten carcasses from two separate flocks processed each day (the second flock on the first shift and the second flock on the second shift) were removed from the line and visually examined for fecal material or ingesta on the inside and outside of the carcass prior to entering the IOBW. Ten carcasses per flock were also evaluated for fecal material or ingesta on the inside or outside of the carcass after exiting the final bird rinse. The number of chickens with contamination was recorded. These data were recorded over 12 days and 24 flocks for a total of 480 carcasses. These data were used to demonstrate whether the process of washing in the IOBW combined with the final bird rinse was having a significant impact on fecal or ingesta contamination on carcasses and thus verifying its efficacy as an intervention for fecal/ingesta contamination.

27.2.6 Online Reprocessing

Most OLR chemicals approved by the USDA-FSIS have been shown to have a significant impact on indicator bacterial levels and *Salmonella* prevalence on ready-to-cook carcasses. Thus, tests were conducted to verify the efficacy of this system.

27.2.6.1 Experimental Procedures

Four carcasses from two separate flocks processed each day (the third flock on the first shift and the third flock on the second shift) were removed from the processing line just prior to the Cecure® system, and four carcasses were collected from the line just after the Cecure system using the aforementioned technique to ensure that no bias was introduced. The carcasses were then individually bagged in sterile polyethylene bags and rinsed using 400 mL of sterile buffered peptone solution. The whole-carcass rinse method was used, and the samples were neutralized as recommended by SafeFoods. The rinsate was encoded as described previously and sent to the laboratory for evaluation for APC, *E. coli* counts, and *Salmonella* prevalence. The microbiological methods used are detailed in the "Microbiological Methods" section of this chapter. There were 192 carcasses evaluated over 12 days and 24 flocks. This allowed the plant to determine the level of APC, *E. coli,* and *Salmonella* on carcasses pre-Cecure and post-Cecure. The numbers of carcasses to be tested were based on statistical models to ensure a representative sampling of the grow-out process.

27.2.7 Redwater Chiller

Regulations in 9 CFR 416.2 (g)(3) state that water that has contacted raw product may be reused for the same purpose or up line provided that measures are taken to reduce physical, chemical, and microbiological contamination or adulteration of the product. This has recently been applied to the use of countercurrent chillers containing water recirculators (redwater systems). Poultry processors must now ensure that they are addressing the physical, chemical, and biological hazards associated with the water that is being removed from the chiller, chilled using a heat exchanger, and reintroduced into the chiller to meet the HACCP regulation requirements. To ensure compliance with this regulation, the plant conducted a test to make sure that the physical, chemical, and biological hazards associated with redwater were addressed, and that these hazards were lower in the water that was reintroduced than they were in the water that was being removed from the chiller to be rechilled by the heat exchanger. In particular, the biological hazard is the hazard of greatest concern. There is a chemical hazard associated with this recirculated water as chlorine is introduced into the water to lower bacterial levels. Thus, the level of chlorine introduced was monitored and was kept at or below 5 ppm free available chlorine at the redwater return. There is no physical hazard associated with recirculated chiller water. While the presence of turbidity or suspended solids does represent a hazard in that it can protect pathogenic bacteria such as *Salmonella* and enable them to survive, it is not a physical hazard. Instead, it contributes to a biological hazard. The hazards that turbidity and suspended solids influence (biological) were assessed in this study.

27.2.7.1 Experimental Procedures

Chiller water samples were collected from the suction box that takes the chiller water to the heat exchanger for chilling. In addition, redwater samples of chilled water returning from the heat exchanger to the chiller were collected. Four samples of each (outgoing and incoming water) were sampled per flock for two separate flocks processed each day (the second flock on the first shift and the second flock on the second shift). This was done for 12 days to represent 24 flocks of chickens. APC, *E. coli* count, and *Salmonella* prevalence were determined for each of these samples. It was hoped that these data would provide evidence that the recirculated water from the chiller had significantly fewer indicator and pathogenic bacteria than the water that was removed from the chiller to be sent to the heat exchanger. In this way, the company could verify that the use of recirculated chiller water was meeting the requirements of 9 CFR 416.2 (g)(3) and that the biological hazard was being reduced. The microbiological methods used for this portion of the study are described in the "Microbiological Methods" section.

27.2.8 Chiller

The chiller is a major intervention step in the reduction of indicator and pathogenic bacteria on broiler carcasses. The chlorine was maintained at or below 5 ppm free available chlorine. The efficacy of the chiller was verified by this study.

27.2.8.1 Experimental Procedures

Four carcasses from two separate flocks processed each day (the third flock on the first shift and the third flock on the second shift) were removed from the processing line just prior to the chiller, and four carcasses were collected from the chiller exit using the aforementioned technique to ensure that no bias was introduced. The carcasses were then individually bagged in sterile polyethylene bags and rinsed using 400 mL of sterile buffered peptone solution containing 0.16 g of sodium thiosulfate as a neutralizing agent by conducting a whole-carcass rinse. The rinsate was encoded as mentioned previously and sent to the laboratory for evaluation for APC, *E. coli* counts, and *Salmonella* prevalence. The microbiological methods used are detailed in the "Microbiological Methods" section of this chapter. There were 192 carcasses evaluated over 12 days and 24 flocks. This allowed the plant to determine the level of APC, *E. coli* counts, and *Salmonella* prevalence on carcasses prechiller and postchiller. The number of carcasses to be tested was based on statistical models to ensure a representative sampling of the grow-out process.

27.2.9 Comparison of Postchill Microbiological Results with Prescald Microbiological Results over Time

This evaluation should enable a processor to answer the most important question of concern to the USDA-FSIS: If the *Salmonella* prevalence on incoming birds (as indicated by prescald values) varies from day to day over time as a result of flock variation, how is the entire collective series of interventions within the plant, including the countercurrent use of water in the scalder and recirculation of water in the chiller, able to control *Salmonella* prevalence on the final product? If the answer to this question is that on days when incoming *Salmonella* levels are very high (80–100%), the finished carcass *Salmonella* prevalence never exceeds 23%, then the combined interventions have been verified to be effective for controlling *Salmonella*. The key to this evaluation is to determine the ability of the plant to reduce *Salmonella*, even in the event of high loads entering the plant.

27.2.9.1 Experimental Procedures

This study was conducted by analyzing data from chickens evaluated prescald and comparing them to carcasses evaluated postchill to determine the average percentage of reduction in *Salmonella* prevalence. APC and *E. coli* were also evaluated to ensure that reductions in these parameters were achieved as well.

27.2.10 Verification that Product Produced during the Startup Phase of the Study Is Safe

In the event that a plant is shut down due to failure to comply with the *Salmonella* performance standard, the USDA-FSIS will require the plant to demonstrate that the product is safe during the startup and testing phase.

27.2.10 Experimental Procedures

Microbiological data from carcasses that were collected postchill were evaluated daily and submitted on a weekly basis to the USDA-FSIS to determine whether the product was safe for consumption. Because the parameters of *E. coli* count and *Salmonella* prevalence are used industrywide, these data were the determining factors to ensure that the product was meeting or exceeding the USDA standards set for raw poultry products.

27.2.11 Long-Term Verification

Once the data from this study have been collected, analyzed, and interpreted, decisions may be made regarding future verification to account for seasonal factors. In the event that very high levels of *Salmonella* are observed coming into the plant, such as 90–100%, and the plant is able to reduce the *Salmonella* prevalence to below 23% on a daily basis, then future verification may not be required because the plant has demonstrated that it can control *Salmonella* under the most adverse conditions. However, if during the study the incoming *Salmonella* numbers are found to be low, then an additional protocol for testing should be submitted to the USDA-FSIS to verify that the processing plant is able to control *Salmonella* prevalence during seasonal variations.

27.3 Microbiological Methods

Microbiological analyses were performed off site by National Calibration and Validation Laboratories, LLC, in El Dorado, Arkansas.

1. APCs were determined using *The Official Methods of Analysis of the AOAC*, Method 990.12, and reported in colony-forming units (CFU).
2. *Escherichia coli:* Tests for *E. coli* were conducted using *The Official Methods of Analysis of the AOAC*, Method No. 990.12, and reported in colony-forming units.
3. *Salmonella: Salmonella* were tested using *The Official Methods of Analysis of the AOAC*, Method No. 2000.07, and reported as either positive or negative.

27.4 Data Analyses

Data were submitted to and statistical evaluation was conducted by the Statistical Consulting Group in the Department of Statistics at the University of Georgia. An additive model was used to compare "pre" and "post" samples separately for each day, but these differences were pooled over all of the test days. This is equivalent to a two-way analysis of variance (ANOVA) with no interaction term. This test is much more accurate because it accounts for the effects due to date. The SAS Analytical Software program was used for the analyses for APC and *E. coli* counts. For *Salmonella* prevalence, logistical regression was conducted using SAS.

27.5 Results and Discussion

The first location of interest is the scalder. The microbiological quality of the carcasses entering and exiting the scalder was evaluated. Microbiological data for APC and *E. coli* counts and *Salmonella* prevalence for carcasses entering and exiting the scalder are presented in Figures 27.1 and 27.2.

These data indicate that APC and *E. coli* counts were reduced, even though the reductions were not practically significant. The statistics indicate that there is a 64% probability that the *Salmonella* prevalence reduction was real as the carcasses traversed the scalder. These data are a clear indication that the scalder was not increasing bacterial levels in any way and was slightly reducing these levels.

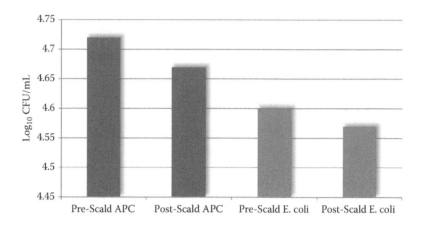

Figure 27.1 **Log$_{10}$ colony-forming units per milliliter APC and *E. coli* on carcasses before and after scalding.**

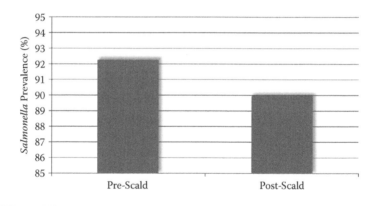

Figure 27.2 *Salmonella* **prevalence on carcasses before and after scalding.**

Results for the microbiological comparison of salvage carcasses after inspection and removal from the line versus those that have been cleaned during the salvage operation are presented in Figures 27.3 and 27.4. These data indicate that the salvage crew was able to significantly reduce the levels of APC and *E. coli* on carcasses, although the numerical reductions were minimal. There was an 89% probability that there was a significant reduction in *Salmonella* prevalence after the salvage carcasses were cleaned. Therefore, it may be concluded that the salvage process is sufficient to reduce bacterial populations on carcasses.

Results for the microbiological comparison of broiler carcasses entering the OLR using cetylpyridinium chloride versus those exiting the OLR system are presented in Figures 27.5 and 27.6. These results indicate that the OLR is

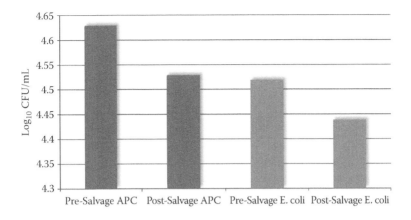

Figure 27.3 Log$_{10}$ colony-forming units per milliliter APC and *E. coli* on carcasses before and after salvage.

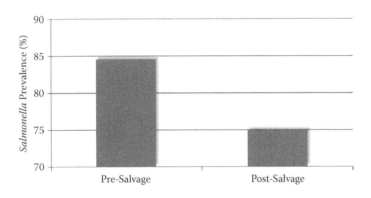

Figure 27.4 *Salmonella* prevalence on carcasses before and after salvage.

doing an excellent job of reducing APC, *E. coli,* and *Salmonella* on carcasses. Numerical reductions (2.4 log$_{10}$ for APC, 2.5 log$_{10}$ for *E. coli*, and a 67.3% reduction in *Salmonella* prevalence) were very high. Statistical analysis revealed that there was greater than a 99.999% chance that the three types of bacteria were being reduced.

Results for the microbiological comparison of carcasses entering the chiller versus those exiting the chiller are presented in Figures 27.7 and 27.8.

APC, *E. coli*, and *Salmonella* prevalence were reduced slightly during chilling. There was a 71% probability and 61% probability that the reductions in APC and *E. coli*, respectively, were due to the chiller treatment. While no significant reductions in *Salmonella* were recorded, there was an overall 1.9% reduction.

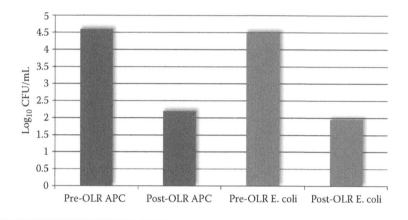

Figure 27.5 **Log₁₀ colony-forming units per milliliter APC and *E. coli* on carcasses before and after OLR.**

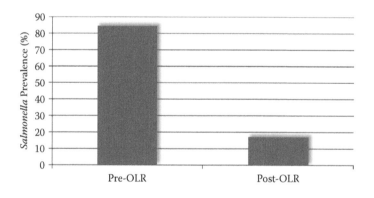

Figure 27.6 *Salmonella* **prevalence on carcasses before and after OLR.**

The overall effect of the interventions in this plant may be determined by testing the microbiological profile of incoming birds and comparing them to carcasses postchill. Overall results for the microbiological comparison of carcasses entering the scalder (postkill) versus those exiting the chiller are presented in Figures 27.9 and 27.10. These data clearly indicate the entire effect of the process on APC, *E. coli,* and *Salmonella.*

These results demonstrate that during processing, APC and *E. coli* were reduced by 2.77 and 2.77 log₁₀, respectively, with a statistical probability of 99.999% chance that the reductions were real. Likewise, *Salmonella* prevalence was reduced from an average of 92.3% coming into the plant to an average of 13.5% exiting the plant postchill over 12 processing days. As with the APC and *E. coli* counts, there was a statistical probability of 99.999% chance that the *Salmonella* reductions were real. These data are promising and verify that the plant was able to control *Salmonella.*

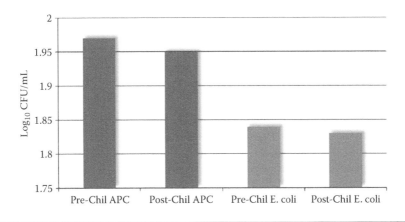

Figure 27.7 Log$_{10}$ colony-forming units per milliliter APC and *E. coli* on carcasses before and after chilling.

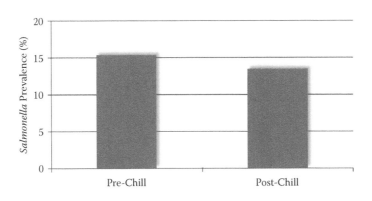

Figure 27.8 *Salmonella* prevalence on carcasses before and after chilling.

27.6 Conclusions

The following conclusions about the individual areas evaluated include

1. The scalder is not increasing bacterial levels in any way and is slightly reducing these levels.
2. APC, *E. coli,* and *Salmonella* were reduced slightly in scald water where the birds exit as opposed to scald water where the birds enter.
3. Salvage significantly reduced APC and *E. coli,* and there was an 89% surety with regard to *Salmonella* prevalence reduction.
4. Cetylpyridinium chloride in the OLR system significantly reduced APC, *E. coli,* and *Salmonella* on carcasses with a surety of 99.999%.

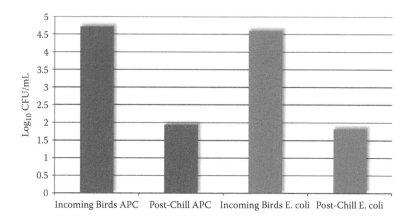

Figure 27.9 Log_{10} **colony-forming units per milliliter APC and** *E. coli* **on carcasses entering and exiting the plant.**

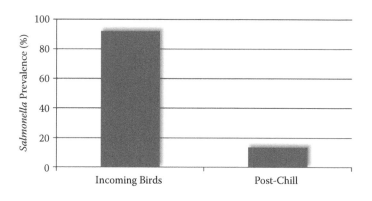

Figure 27.10 *Salmonella* **prevalence on carcasses entering and exiting the plant.**

5. Chilling had little effect on APC and *E. coli* but did decrease *Salmonella* prevalence by almost 2% and did not increase any of these parameters.
6. Overall reduction of APC, *E. coli*, and *Salmonella*, when taking the entire processing operation into account, was dramatic and significant.

Collectively, these results indicate that the poultry-processing plant evaluated was able to verify adequately that these bacterial populations were being controlled throughout processing.

References

Musgrove M T, Cox N A, Bailey J S, Stern N J, Cason J A, and Fletcher D L (1997), Effect of cloacal plugging on microbial recovery from partially processed broilers, *Poultry Science*, 76, 530–533.

Russell S M (2005), Verification of intervention strategies for reducing pathogenic and indicator bacteria on broiler carcasses: A case study, *Poultry USA Magazine*, December, 40, 42, 44, 46, 48, 50, 52, 54–56, 58, 60, 61.

Chapter 28

Salmonella Intervention Strategies and Testing Methods Differ Greatly between the United States and Europe: What Are the Implications?

28.1 Introduction

The United States and Europe have evolved different approaches for controlling *Salmonella* on raw poultry products. In addition, the methods used to test poultry products for the presence of *Salmonella* vary greatly from nation to nation.

28.2 Production Differences

In the United States, companies are limited regarding the types of interventions they may use to control *Salmonella* during the processes of breeding, hatching, and grow out. The reasons for these limitations are many, including economic and environmental factors, regulatory agency restrictions, and the massive scale of production (Russell et al., 2009). In Europe, there is an intense fear of using chemicals to eliminate *Salmonella* on chicken products. Consumers are much less accepting of the use of any chemical intervention during processing. As such, no chemicals are approved for use in poultry-processing facilities in Europe.

Therefore, a great deal of emphasis is placed on interventions during breeding, hatching, and grow out. For example, in Europe, some countries test all breeder flocks for *Salmonella,* and if a flock is found to be positive for *Salmonella*, the company destroys the entire breeder flock. Most studies show that, using this extreme measure, these countries have been able to significantly reduce *Salmonella* to only 1–6% on birds coming into the processing facility. One might ask why the United States has not implemented such extreme measures. This approach is impossible in the United States because, for example, we produce twice as much poultry in Athens, Georgia, than is produced in all of Sweden, where these practices are common. The scale of production in the United States makes this approach absolutely impractical.

Another approach in some of the E.U. countries that has been used for many years is known as competitive exclusion. In Europe, adult chickens that are found to be free of *Salmonella* are thought to have competitive bacterial flora in their intestines that prevent colonization of the chicken with *Salmonella*. These chickens are euthanized, and their intestinal tracts are removed. The lining of the intestines are scraped into a container (mucosal scrapings), and these bacteria are then grown in large containers to very high numbers. This solution is then sprayed onto the baby chicks to allow for colonization of their intestines with "good bacteria." These bacteria then colonize the baby chick's intestines and prevent *Salmonella* from attaching. Also, these bacteria produce compounds that kill *Salmonella*. This approach is known as *undefined competitive exclusion* (because the bacteria in the mixture have not been identified and are unknown). This approach is illegal in the United States because the U.S. Food and Drug Administration (FDA) requires that all bacteria that are fed to baby chicks must be identified and characterized to ensure that none of them is pathogenic and none of them is antibiotic resistant. Because of this requirement, no commercially available undefined competitive exclusion products are allowed to be sold for *Salmonella* control during grow out.

The main approach in the United States to reduce *Salmonella* during grow out involves the use of vaccines. The vaccines are variable in terms of their efficacy because there are greater than 2,500 serotypes of *Salmonella,* and vaccines cannot be made to prevent all of them. There are five to seven main serotypes of *Salmonella* commonly isolated from poultry carcasses in the United States. The top five in order of importance are (1) Kentucky, (2) Enteritidis, (3) Heidelberg, (4) Typhimurium (var. Copenhagen), and (5) Typhimurium (U.S. Department of Agriculture, Food Safety Inspection Service [USDA-FSIS], 2007) and are presented in Figure 28.1.

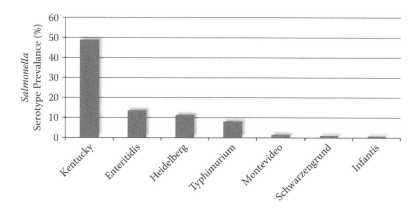

Figure 28.1 *Salmonella* **prevalence by serotype most commonly found by the USDA-FSIS on broiler chickens in the United States.**

Roy et al. (2005) reported isolating 569 *Salmonella* from 4,745 samples from poultry products, poultry, and poultry environments in 1999 and 2000 from the Pacific Northwest. These *Salmonella* were identified to their exact source, and some were serogrouped, serotyped, phage typed, and tested for antibiotic sensitivity. Food product samples tested included rinse water of spent hens, broilers, and chicken ground meat. Poultry environment samples were hatchery fluff from the hatcheries where eggs of grandparent broiler breeders or parent broiler breeder eggs were hatched and drag swabs from poultry houses. Samples of liver or yolk sac contents collected at necropsy from the young chicks were received in the laboratory. Of these samples tested, 569 were *Salmonella* positive (11.99%). Ninety-two *Salmonella* were serogrouped with polyvalent somatic antisera A-I and the polymerase chain reaction. Somatic serogroups B and C comprised 95.25% of all the *Salmonella*. Of a total of 569 positive samples, 97 isolates of *Salmonella* were serotyped. A total of 16 serotypes and an unnamed *Salmonella* belonging to serogroup C1 were identified. The *Salmonella* serotypes were *heidelberg* (25.77%); *kentucky* (21.64%); *montevideo* (11.34%); *hadar* and *enteritidis* (5.15% each); *infantis, typhimurium, ohio*, and *thompson* (4.12% each); *mbandaka* and *cerro* (3.09% each); *senftenberg* (2.06%); *berta, istanbul, indiana*, and *saintpaul* (1.03% each); and an unnamed monomorphic *Salmonella* (2.06%).

In the European Union, the serotypes of most concern are (1) *enteritidis*, (2) *typhimurium*, (3) *hadar*, (4) *infantis*, and (5) *virchow*. Thus, vaccines created in the European Union directed against their serotypes of concern would have little effect in the United States. This is because, if a vaccine is directed against one or two of these serotypes, it may not be effective if the other serotypes are found in the environment of the chicken during grow out. Companies are doing a much better job of making the vaccines effective against a broad variety of serotypes. Even when using these vaccines, reductions of only 50% in *Salmonella* prevalence on incoming broilers is common. Thus, additional measures in the field must be taken to decrease *Salmonella*.

Another measure used by some companies is the addition of an acid or acid blend to watering systems during the feed withdrawal phase before slaughter. This disinfects the crops of the birds and decreases *Salmonella* in the crops that were consumed by chickens during the feed withdrawal period. Because of these considerations and the cost of intervention before slaughter is higher than using chemicals in the plant, most emphasis in the United States is placed on the slaughter operation.

28.3 Processing Differences

28.3.1 Picking

Nde et al. (2007) reported that scald water and the fingers of the picker machines may contribute to the contamination of *Salmonella*-free flocks when they are processed following a *Salmonella*-positive flock. The results of this study showed evidence for the possible transfer of *Salmonella* from turkey feathers to carcass skin during defeathering. This direct contaminating effect was greater during Visits 1 and 4, when the *Salmonella* prevalence after defeathering increased significantly. *Salmonella* contamination on turkey feathers may therefore be a useful indicator of the potential for cross contamination during defeathering. Future strategies could focus on reducing the level of *Salmonella* on the feathers of live birds, thus minimizing the risk of cross contamination during defeathering. This is important because in plants in the European Union, if no chemical interventions are used, then *Salmonella* prevalence will surely increase during picking, regardless of how well the company is able to decrease *Salmonella* in the live birds, unless the prevalence is absolutely zero.

28.3.2 Differences in Perspective

The USDA-FSIS views *Salmonella* on poultry as a food safety issue and regulates the prevalence of *Salmonella* that is allowable on poultry carcasses. However, the European Union does not view *Salmonella* as a food safety issue but as a sanitation indicator (Cox et al., 2009). In the United States, carcasses are sampled at the end of the chiller by the USDA-FSIS, and the samples are evaluated for *Salmonella*. If samples are in excess of the *Salmonella* performance standard (13 positive carcasses out of 51), then the USDA penalizes the plant. If this occurs three times, inspection is withdrawn, and the plant is shut down. However, in Europe, no such regulations exist. No penalties are levied due to excessive *Salmonella* prevalence on carcasses.

Figure 28.2 Typical immersion chiller used in the United States.

28.3.3 Chilling

In the United States, over 99% of companies use water immersion chilling systems (Figure 28.2). In Europe, air chilling (Figure 28.3) is most commonly used. This is important because a properly run immersion chiller is the most effective intervention tool available for poultry processors. Many companies in the United States are able to maintain *Salmonella* at very low levels on carcasses using the chiller as the main intervention strategy. In Europe, no chemicals, including chlorine, are used to reduce *Salmonella* during processing. This begs the question of what happens when a flock that is contaminated with *Salmonella* enters the plant or when the interventions used in the field break down. In a word, nothing. The European Union does not have any *Salmonella* regulations for ready-to-cook poultry carcasses.

28.3.4 Sampling

The way that poultry is sampled and tested varies greatly depending on the country where the tests are done. In the United States, the USDA-FSIS inspectors rinse a chicken with 400 mL of sterile buffered peptone water (whole-carcass rinse). However, in the European Union, plant employees collect a 25-g neck skin sample from three different carcasses and pool them. Cox et al. (2009) conducted a study to determine which method is most sensitive for detecting the presence of *Salmonella* on carcasses. The research demonstrated that both methods are fairly equivalent for

Figure 28.3   Typical air chiller used in the European Union.

detecting *Salmonella,* but that neither is sensitive enough to be considered perfect. For example, on many carcasses, the neck skin method picked up the *Salmonella,* but none was found in the carcass rinse for that carcass; in other cases, the reverse occurred (Figures 28.4 and 28.5).

Based on this study, both methods would need to be used together to really obtain a good idea of actual prevalence. It is important to note that in some countries around the world, particularly for exported product, the test method used is completely different from the two methods used regularly in the United States and European Union. The chicken skin is sterilized using a blowtorch or iodine solution, then the skin is removed using sterile tweezers, and a sample of deep breast muscle is taken and tested for *Salmonella.* Amazingly, *Salmonella* is never found using this technique, allowing the company/country to state boldly that it does not have any *Salmonella* on its poultry. This is misleading and causes great confusion. By this testing method, a company could say that their chicken are sterile, which is of course ridiculous. Meanwhile, the USDA-FSIS is forcing companies in the United States to post their *Salmonella* prevalence and company names, addresses, and plant number (P-numbers) on the Internet for the world to see. This causes a potential imbalance in trade based on completely false data.

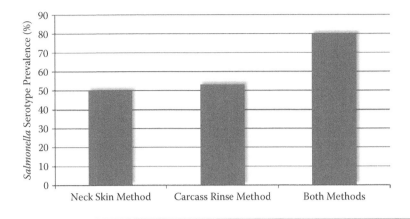

Figure 28.4 *Salmonella* **prevalence on carcasses of 30 tested for each method taken from carcasses before the inside-outside bird washer.**

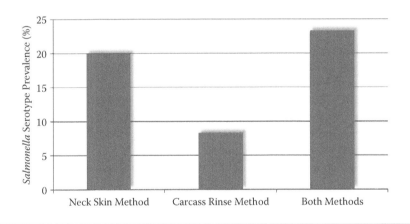

Figure 28.5 *Salmonella* **prevalence on carcasses of 30 tested for each method taken from carcasses after chilling.**

28.4 Implications

Poultry companies in this country are placed in a very difficult situation. They are required to use chemicals in the plants to lower *Salmonella* to acceptable levels for the USDA. They do an excellent job in this regard. However, because they use chemicals, they cannot export to Europe. Moreover, they cannot use cost-effective measures to control *Salmonella* during grow out because they are too expensive or

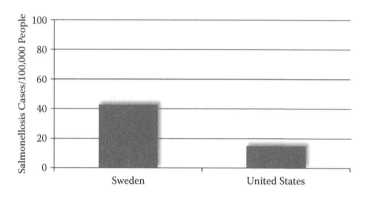

Figure 28.6 Salmonellosis cases per 100,000 people in Sweden (where extensive expensive efforts are made to eliminate *Salmonella* in live birds) and the United States.

are against the law due to FDA regulations. Even though they effectively lower *Salmonella* to 7.5% nationwide on postchill carcasses, this is not acceptable to countries that have a "zero tolerance" for *Salmonella* regulation for imported poultry, which is based on an absurd testing method. To add to the difficulty, now the companies that are in Category 2 or 3 of the *Salmonella* performance standard must have their *Salmonella* data posted on the Internet, which eliminates their exportation to zero-tolerance countries.

An extremely important question that must be answered is what are the Europeans getting for the incredible expenditure of effort and money to eliminate *Salmonella* from the breeders, hatchery, and grow-out operation. For example, what impact does this effort have on human salmonellosis?

Information from Cox et al. (2009) tells the story (Figure 28.6). This information demonstrated that salmonellosis is far greater (42.8 per 100,000 people vs. 14.9 per 100,000 people in the United States) in a country where extraordinarily expensive measures are used to eliminate *Salmonella* from the flocks prior to processing.

There should be an effort by leaders of these countries to use sound scientific principles to come together and agree on compatible methods for eliminating and testing for *Salmonella*. There is no logical reason why a method used in the field for many years in Europe to eliminate *Salmonella* from the flock (competitive exclusion) without any adverse effects cannot be used in the United States. This causes great confusion for companies that operate globally and for consumers who believe they are buying "*Salmonella*-free" chicken.

References

Cox N A, Richardson L J, Cason J A, Buhr R J, Smith D P, Cray P F, and Doyle M P (2009), Comparison of neck skin versus whole carcass rinse for prevalence of *Salmonella* and *E. coli* counts recovered from broiler carcasses, Presented at the U.S. Poultry and Egg Association Exposition, Atlanta, GA (January 2009).

Nde C W, McEvoy J M, Sherwood J S, and Logue C M (2007), Cross-contamination of turkey carcasses by *Salmonella* species during defeathering, *Poultry Science*, 86, 162–167.

Roy P, Dhillon A S, Lauerman L H, Schaberg D M, Bandli D, and Johnson S (2005), Results of *Salmonella* isolation from poultry products, poultry, poultry environment, and other characteristics, *Avian Diseases*, 46, 17–24.

Russell S M, Cox N A, and Richardson J L (2009), Zero tolerance for *Salmonella* raises questions, *Poultry USA Magazine*, April, 20, 21.

U.S. Department of Agriculture, Food Safety Inspection Service (2007), Serotypes profile of *Salmonella* isolates from meat and poultry products: January 1998 through December 2007. http://www.fsis.usda.gov/Science/Serotypes_Profile_Salmonella_Isolates/index.asp.

Chapter 29

Impact of the New USDA-FSIS *Salmonella/Campylobacter* Performance Standards for Young Chickens

29.1 Introduction

The U.S. Department of Agriculture, Food Safety Inspection Service (USDA-FSIS), announced new regulations implemented in July 2011 in which broiler chickens produced in the United States will be subject to new performance standards for *Salmonella* and a performance standard for *Campylobacter* for the first time (Russell, 2011). In this chapter, the new *Salmonella* performance standards for the poultry industry and its potential impact are evaluated.

29.2 Estimated Decrease in Food-Borne Illness

The USDA-FSIS has estimated that, once the new standard is in place for 2 years, 39,000 illnesses due to *Campylobacter* and 26,000 illnesses due to *Salmonella* will be

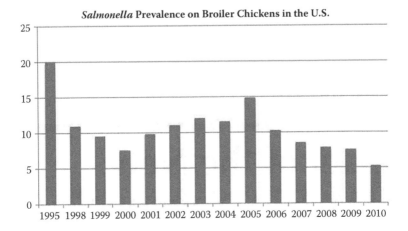

Figure 29.1 *Salmonella* **prevalence on broiler chickens in the United States from 1995 to 2010.**

eliminated. This is in conflict with currently available data regarding human food-borne illness due to *Salmonella,* which the Centers for Disease Control and Prevention (CDC) reports has flatlined for 20 years (see Figures 29.1–29.3). If consumption of poultry were contributing to the prevalence of salmonellosis, then changes in prevalence on fresh poultry would be reflected in changes in human salmonellosis. This relationship has not been established. In fact, *Salmonella* prevalence increased from 2000 to 2005 and decreased dramatically from 2005 to 2010, but no significant changes in human salmonellosis occurred. This clearly indicates that consumption of poultry is not having a significant impact on salmonellosis in humans. The question is what impact on human salmonellosis will occur if the industry spends millions, if not billions, of dollars meeting these new requirements. Is it worth it? Was the billion dollars spent to implement the Hazard Analysis and Critical Control Point Final Rule/*Salmonella* performance standard worth it? What effect did it have on human salmonellosis? There needs to be government accountability for arbitrary implementation of regulation without any and, in some cases, contradictory evidence that the new regulation will have an impact on food-borne illness.

When the limit was set at 20% for *Salmonella,* in the 1996 USDA-FSIS *Salmonella* Performance Standard, numerous interventions existed in the field and in the plant that could be implemented to assist the industry in meeting this standard. In so doing, the industry was able to lower the numbers, on average to 7.5% by 2000. However, retailers of poultry products were receiving continuous pressure by consumer groups to use chicken that had not received growth-promoting antibiotics. The industry responded after 2000 by drastically reducing the use of growth-promoting antibiotics. Also, during that time, the industry was trying to reduce water usage to be more environmentally friendly. Many plants went from using 8 gallons/bird to as low as 4 gallons/bird. The washing and dilution effect

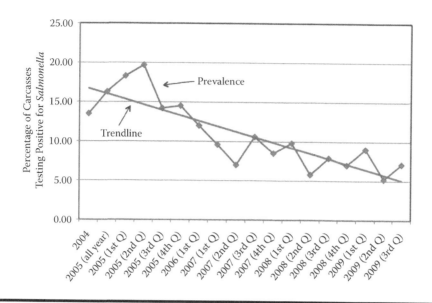

Figure 29.2 *Salmonella* **prevalence trend on broiler chickens in the United States from 2004 to 2009.**

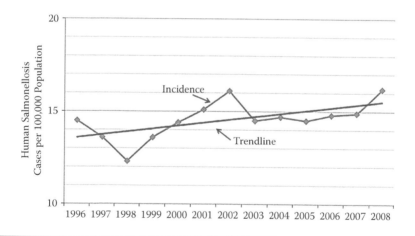

Figure 29.3 **Human salmonellosis cases per 100,000 people from 1996 to 2008 in the United States.**

of the added water use at 8 gallons/bird provided a cleaner product. Both of these factors likely had an impact on *Salmonella* prevalence on chickens. There is a misconception that the industry was intentionally discontinuing its intervention strategies and cutting back on chemical usage. This is simply not true. I traveled to plants throughout the United States during that time and assisted them with issues regarding meeting the performance standards. Moreover, there seems to be a misconception that when the USDA-FSIS threatened to strengthen the regulations in 2005, suddenly the *Salmonella* prevalence decreased. This relationship is false. There was a significant change in the way that the industry was allowed to use and monitor chlorine usage. This single factor had more impact on *Salmonella* prevalence than any other I have observed over the years. The idea that, because USDA-FSIS puts pressure on the industry to improve, it improves and when the USDA-FSIS does not put pressure on it, *Salmonella* prevalence increases is not an accurate assessment.

29.3 The New *Salmonella* Performance Standard Fails to Take into Account Two Important Considerations

29.3.1 *Variables Outside the Control of the Poultry Industry that Have an Impact on* Salmonella *Prevalence*

Seasonality, humidity, and disease conditions all have a significant impact on *Salmonella* prevalence on poultry. These factors cannot be controlled by the industry. Studies have demonstrated that *Salmonella* prevalence varies by season. Moreover, articles have detailed how higher relative humidity in poultry houses can increase prevalence. I authored an article in *Poultry Science* (Russell, 2003) that detailed how air sacculitis infections increase *Campylobacter* counts on broilers and in another popular article detailed analysis of over 5 million chickens; *Salmonella* prevalence was significantly higher on birds with air sacculitis as well. Unfortunately, the FDA decided to ban the use of the only antibiotic (Enrofloxacin) that is effective for treating air sacculitis (as shown in a study by the Poultry Diagnostic Research Center at the University of Georgia at Athens). With these variables, a plant may be using a particular set of interventions day in and day out with great success, but if the temperature, humidity, or disease outlook changes, the plant may not be able to meet the standard. How can the growers affect the weather? How can they control diseases when their only effective tools have been taken away? This seems patently unfair.

29.3.2 *Sampling Methodologies Place the Poultry Industry at a Competitive Disadvantage*

The way that poultry is sampled and tested varies greatly depending on the country where the tests are done. In the United States, the USDA-FSIS inspectors rinse a chicken with 400 mL of sterile buffered peptone water (whole-carcass rinse). However, in the European Union, plant employees collect a 25-g neck skin sample from three different carcasses and pool them. As mentioned in Chapter 28, Cox et al. (2009) of the USDA Agricultural Research Service (ARS) conducted a study to determine which method is most sensitive for detecting the presence of *Salmonella* on carcasses. The research demonstrated that both methods are fairly equivalent for detecting *Salmonella,* but that neither is sensitive enough to be considered perfect. For example, on many carcasses, the neck skin method picked up the *Salmonella,* but none was found in the carcass rinse for that carcass; in other cases, the reverse occurred.

Based on this study, both methods would need to be used together to really obtain a good idea of actual prevalence. This also highlights the differences in methods used for sampling turkeys. The sampling method used for chickens uses a whole-bird carcass rinse by which the entire surface area (both inside and out) of the chicken is sampled. Since the inside of the carcass is more likely to contain cross contamination from the evisceration process, the likelihood of detecting *Salmonella* and *Campylobacter* is greatly increased even though their numbers were reduced through processing. In contrast, the sampling method used for turkeys involves taking small skin swabs (50 cm² each) on the thigh and back of one-half of the turkey carcass. This method is biased against finding *Salmonella* and *Campylobacter,* which are most likely to be found on other areas of the carcass. The FSIS provides no justification for subjecting broiler carcasses to different testing methods and standards. Efforts should be made by the USDA-FSIS to standardize these methods or keep the data obtained from processors in the United States confidential.

29.4 *Campylobacter* Performance Standards

A *Campylobacter* standard puts the poultry industry in a difficult position. Currently, there are no intervention methods that have been demonstrated to be consistently effective for eliminating *Campylobacter* in chickens during grow out. Interventions used to control *Salmonella* transmission from breeder to baby chick (vaccination and hatchery intervention) and to prevent colonization of baby chicks (vaccination and competitive exclusion) have proven ineffective for *Campylobacter*. In fact, the scientific community is still divided on whether *Campylobacter* is vertically transmitted. If we do not fully understand how the organism is colonizing baby chicks, how do we implement effective interventions? Thus, the industry is left with no tools for controlling *Campylobacter* in broiler populations. This means that all interventions must be implemented at the plant level. As with grow out, no scientific studies exist that demonstrate that one particular intervention works well for controlling *Campylobacter*.

29.4.1 *Microbiological Testing for* Campylobacter *Presents a Burden to the Industry*

The reason why *Campylobacter* has been getting attention over the last 10 years has nothing to do with this organism being an "emerged pathogen" as with *Escherichia coli* O157:H7. *Campylobacter* was described in 1886 by Theodor Escherich of *Escherichia coli* fame. This begs the question of why we are focusing on it only now. The answer is because it is hard to detect microbiologically. We have not had widely available tools for many years to culture this organism. As such, making the poultry industry test for this organism will require specific and expensive CO_2 incubators, specialized media, a phase contrast microscope, and someone with far more microbiological training than most poultry companies currently employ. Thus, it will cost the poultry industry much more money to test products for this organism.

29.5 FSIS Does Not Have Statutory Authority to Regulate Nonadulterant Pathogens

Under the Fifth Circuit's decision in *Supreme Beef Processors, Inc. v. USDA*, the FSIS *Salmonella* performance standard regulations fall outside the statutory grant of rule-making authority. The court held that the USDA has no statutory authority to regulate the levels of nonadulterant pathogens, and that *Salmonella* is not an adulterant per se in raw poultry. Thus, the USDA lacks legal authority to regulate *Salmonella* levels in end products. The FSIS has stated that the performance standards are used as targets without regulatory consequences; however, in reality, the agency uses *Salmonella* performance standard failures to justify increased frequency of monitoring, implementation of comprehensive food safety assessments, notices of intended enforcement actions, and the posting of establishment names on the FSIS Web site. Moreover, establishments whose *Salmonella* prevalence is posted on the Internet would have difficulty exporting their poultry, particularly to countries that use incorrect testing methods to claim zero *Salmonella* prevalence. Therefore, there are clear and concrete regulatory and commercial consequences resulting from these regulations.

29.6 The Administrative Procedure Act

When proposing new laws, it is essential that the USDA-FSIS follow the law regarding rule making. The law requires that the FSIS must engage in a notice-and-comment period to allow interested parties the opportunity to present evidence supporting or in opposition to the new rule. The performance standards recently announced by FSIS constitute a legislative rule under the Administrative Procedure Act because they impose obligations and significant effects on private interests and are binding on private parties; however, no comment period was provided as necessary by law. In addition, the FSIS also is required to inform the public concerning the performance standards through notice-and-comment rule making because it is effectively amending an existing regulation that was put forth, whether or not the Fifth Circuit in effect found the existing performance standards to be without legal basis. Without this public comment period, all available scientific data and other relevant information, including economic impact, have not been thoroughly evaluated, bringing into question whether the resulting standards are scientifically sound and feasible. This is important because without this comment period, scientific data that clearly demonstrate no relationship between human salmonellosis and prevalence of *Salmonella* on poultry were not publicly acknowledged. It is really important for the public to be aware when regulations are enacted without understanding the economic impact they will have on the industry and the cost of food.

The new performance standards will only place additional burdens on the industry, possibly resulting in higher food costs to the consumer with no measurable public health benefit. Such significant changes without strong scientific data showing public health benefits place the U.S. chicken industry at a competitive disadvantage for export opportunities.

References

Cox N A, Richardson L J, Cason J A, Buhr R J, Smith D P, Cray P F, and Doyle M P (2009), Comparison of neck skin versus whole carcass rinse for prevalence of *Salmonella* and *E. coli* counts recovered from broiler carcasses, Presented at the U.S. Poultry and Egg Association Exposition, Atlanta, GA, January 2009.

Russell S M (2003), The effect of air sacculitis on bird weights, uniformity, fecal contamination, processing errors, and populations of *Campylobacter* spp. and *Escherichia coli*, *Poultry Science*, 82, 1326–1331.

Russell S M (2011), New *Salmonella*, *Campylobacter* performance standards for poultry lack scientific foundation, *Poultry USA*, June, 18, 20, 21.

Chapter 30

Future Outlook

30.1 Introduction

Poultry will always be a desirable, inexpensive, healthy source of protein and will only grow in popularity in the future. However, the pressures placed on the poultry industry by poor governmental policies (such as those that influence ethanol production), implementation of policies as the result of pressure by animal rights organizations, and pressure to produce safer food by the U.S. Department of Agriculture, Food Safety Inspection Service (USDA-FSIS), will make it more expensive to produce poultry in the future. The reality, based on available data, is that the United States produces the safest poultry in the world (Figure 30.1).

This causes one to ask, why continually lower the standard? What is the desired effect? Can *Salmonella* be completely eliminated from poultry? The answers are detailed next.

30.2 The Law of Diminishing Return

There is a tenant in food microbiology that when disinfecting a food item, much more effort is required to eliminate the last few organisms than is required to eliminate large numbers of bacteria. In fact, 90% of the effort is required to eliminate the last 10% of bacteria. This is because the first 99% are low-hanging fruit. These are bacteria that are not encased in biofilms, are loosely attached, and are in a location where the chemical can reach them. The remaining few organisms are firmly attached, encased in biofilms, and cannot be disinfected easily. The USDA-FSIS,

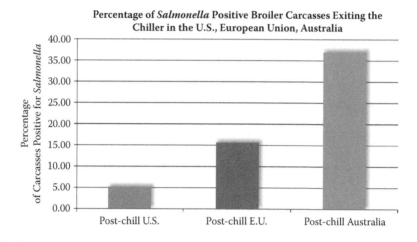

Figure 30.1 Percentage of *Salmonella***-positive broiler carcasses exiting the chiller (going into the marketplace) in the United States, the European Union, and Australia.**

by making the new standard less than 7.5%, is essentially saying that they want 92.5% of the chickens to have zero *Salmonella*. Stated another way, the USDA-FSIS is saying that they want the industry to eliminate the last few remaining *Salmonella* cells on many more chickens than they required previously. This is a tall order as these cells are firmly attached and require much more effort to eliminate. The USDA-FSIS may be expecting the industry to turn some knobs and make adjustments in response to the new the regulation that will make the prevalence decrease further. The reality is that the industry will have a very hard time eliminating those last few *Salmonella* cells that are making some of the chickens positive for *Salmonella*.

The overall outlook for the poultry industry is very good because of the continual increasing demand for poultry. However, the industry faces outside pressures that make it more expensive and difficult to produce and sell their products.

Index